A Level
Mathematics
for Edexcel

# Mechanics

M1

Brian Jefferson

OXFORD
UNIVERSITY PRESS

# OXFORD
## UNIVERSITY PRESS

Great Clarendon Street, Oxford OX2 6DP

Oxford University Press is a department of the University of Oxford.
It furthers the University's objective of excellence in research, scholarship,
and education by publishing worldwide in

Oxford   New York

Auckland   Cape Town   Dar es Salaam   Hong Kong   Karachi
Kuala Lumpur   Madrid   Melbourne   Mexico City   Nairobi
New Delhi   Shanghai   Taipei   Toronto

With offices in

Argentina   Austria   Brazil   Chile   Czech Republic   France   Greece
Guatemala   Hungary   Italy   Japan   South Korea   Poland   Portugal
Singapore   Switzerland   Thailand   Turkey   Ukraine   Vietnam

Oxford is a registered trade mark of Oxford University Press
in the UK and in certain other countries

British Library Cataloguing in Publication Data

Data available

ISBN 9780-19-911781 9

10 9 8 7 6

Printed in Great Britain by Ashford Colour Press Ltd, Gosport.

Paper used in the production of this book is a natural, recyclable product
made from wood grown in sustainable forests. The manufacturing process
conforms to the environmental regulations of the country of origin.

**Acknowledgements**

The photograph on the cover is reproduced courtesy of Photos.com/Jupiter Images

The Publisher would like to thank the followig for permission to reproduce
photographs:
**P6** David Luscombe/iStockphoto; **P42** Chris McKenna/Wikimedia; **P98** Eric
Renard/iStockphoto; **P126** Alex Gore/Alamy; **P146** Jan Rihak/iStockphoto;
**P166** Photodisc.

The publisher would also like to thank Mike Heylings and Kathleen Austin
for their expert help in compiling this book.

# About this book

Endorsed by Edexcel, this book is designed to help you achieve your best possible grade in Edexcel GCE Mathematics Mechanics 1 unit.

Each chapter starts with a list of objectives and a 'Before you start' section to check that you are fully prepared. Chapters are structured into manageable sections, and there are certain features to look out for within each section:

Key points are highlighted in a blue panel.

Key words are highlighted in bold blue type.

Worked examples demonstrate the key skills and techniques you need to develop. These are shown in boxes and include prompts to guide you through the solutions.

> Derivations and additional information are shown in a panel.

Helpful hints are included as blue margin notes and sometimes as blue type within the main text.

Misconceptions are shown in the right margin to help you avoid making common mistakes.

Investigational hints to prompt you to explore a concept further.

Each section includes an exercise with progressive questions, starting with basic practice and developing in difficulty. Some exercises also include 'stretch and challenge' questions marked with a stretch symbol ·

At the end of each chapter there is a 'Review' section which includes exam style questions as well as past exam paper questions. There are also two 'Revision' sections per unit which contain questions spanning a range of topics to give you plenty of realistic exam practice.

The final page of each chapter gives a summary of the key points, fully cross-referenced to aid revision. Also, a 'Links' feature provides an engaging insight into how the mathematics you are studying is relevant to real life.

At the end of the book you will find full solutions, a key word glossary and an index.

---

EXAMPLE 1

Points $A$ and $B$ have position vectors $a = 2i + j$ and $b = 5i - 6j$ respectively. Find the distance $AB$.

$\overrightarrow{AB} = b - a$
$= (5i - 6j) - (2i + j) = 3i - 7j$

Find the magnitude:

$AB = |b - a|$
$= \sqrt{3^2 + (-7)^2} = \sqrt{58} = 7.62$

# Contents

**M1**

# 1

# Modelling

This chapter will enable you to
- understand how the modelling process is used when mathematics is applied to 'real-world' problems
- use conventional terms to show what assumptions are being made.

## Introduction

You can think of mathematics as falling into two categories – pure and applied.

Pure mathematics has no immediately obvious physical application. However, the ideas and techniques you learn there turn out to be vital when mathematics is applied to solving real-world problems.

e.g. The techniques you need for mechanics are those of basic algebra and trigonometry.

Mechanics is the branch of applied mathematics in which you attempt to understand and predict the behaviour of physical objects under the action of forces.
In a given situation you need to filter out those factors which will have little effect on the outcome, and base your equations on just the major factors.

e.g. If you are dragging a box along the ground, friction would be a major factor, but air resistance would play no significant part in what happens.

This process of simplifying the situation is known as mathematical modelling. It is a process common to all applications of mathematics. Whether you are trying to predict the path of a hurricane or the effect of a change in the financial interest rate, you can only ever work with a simplified picture of the situation.

The real world is complicated. Exploring a situation mathematically is only possible if you simplify things. To do this you must make assumptions about which factors to include in your calculations and which factors can be ignored. Equations based on this simplified setup are a *mathematical model* of the situation. The predictions from this model can be compared with the real-world outcomes and, if necessary, the assumptions can be modified to give a better model.

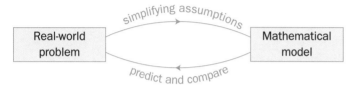

**Consider the following situation.**
A group of people plan to drive directly across a stretch of desert from their present position *A* to a campsite at *B*. They want to predict the length of the journey. They have a map of the region with a scale of 1 cm : 1 km.

The group follow stages when modelling their journey:

**The model.**
Their proposed model is:

> *A straight line AB drawn on the map is a scale model of the journey.*

first stage: define the model and decide what assumptions to make;

**The assumptions.**

1 The journey is 'flat'. Extra distance caused by hills is insignificant compared with the length of the journey. The model tends to under-estimate the actual distance driven.

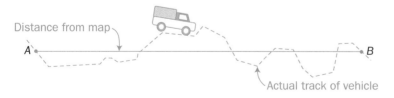

2 The journey is in an exact straight line. In practice there are probably rocks and other obstacles to go round. The model is again likely to produce an under-estimate.

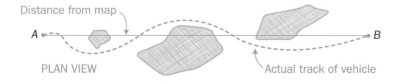

MI

3  The journey is short. Distortions caused by the fact that the
   map is flat and the Earth's surface is curved are not noticeable.

**Prediction and comparison**

Suppose that the straight line *AB* on the map is 18.6 cm.
The model predicts that the journey will be 18.6 km.

When they arrive at *B*, the group checks the distance travelled and
finds it is 19.2 km.

When comparing the model with reality they must consider
sources of error, both in the prediction and in the measurement
of the actual distance.

*second stage: obtain predicted
results from the model;*

*third stage: obtain practical data.
The two sets of results can then
be compared;*

1  The measurement of *AB* on the map is only accurate to the
   nearest millimetre. In addition, identifying the start and finish
   points on the map can only be approximate. The error bounds
   could perhaps be

   18.5 km ⩽ predicted distance < 18.7 km.

2  The actual distance would be found using the milometer on
   their vehicle. This shows the nearest completed 0.1 km, so the
   journey has error bounds

   19.2 km ⩽ actual distance < 19.3 km.

**Evaluating the model**

Allowing for the above errors, the group could decide if the
model was accurate enough for their purpose. If not, new
assumptions and a modified model would be needed for their
next journey. They might, for example, obtain a larger scale map
and measure a route including detours around likely obstacles.

*final stage: decide whether the
model is a good one. If it is not
good enough, the assumptions
must be reconsidered and a
modified model developed.*

## The modelling process

All applications of mathematics to real-world problems follow
this process, which can be summarised in a flowchart.

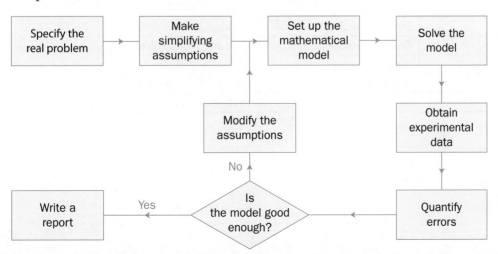

MI

EXAMPLE 1

Objects of mass 3 kg and 4 kg are connected together by a rope which passes over a fixed pulley. The objects are held still, with the rope sections straight and vertical, and then released. What simplifying assumptions could be made so that the motion could be calculated?

**What factors might affect things?**
An (incomplete) list might include:
- the mass, size and shape of the objects
- the length, mass and elasticity of the rope
- the size, mass and friction of the pulley
- gravity
- air resistance.

**Which factors can be ignored?**
Only gravity and the mass of the objects must definitely appear in the model. The likely assumptions are:
- the size of the objects is small enough to be ignored. They can each be treated as a mass concentrated at a single point
- the mass of the rope is insignificant compared with that of the objects
- the rope does not stretch significantly
- the pulley is light
- frictional forces at the pulley are small enough to ignore
- air resistance is negligible.

The object is described as **small** or as being a **particle**.

The rope is described as **light**.

The rope is described as **inextensible**.

The pulley is described as **smooth**.

## Conventional terms
Certain words are used to indicate assumptions being made.
This table shows some of the most common terms.

| Term | Applied to | Meaning |
|---|---|---|
| Light | Strings, springs, rods, pulleys | Mass is negligible |
| Small, or a Particle | Objects | Size is negligible |
| Inextensible | Strings, rods | Does not stretch |
| Rigid | Beams, objects, structures | Does not bend or deform |
| A rod | Beams, struts | Rigid and of negligible width |
| Smooth | Surfaces, pulleys | Friction is negligible |
| Rough | Surfaces | Friction cannot be ignored |
| A lamina | Flat sheets of metal etc. | Thickness is negligible |

## Exercise 1.1

1 Do you think air resistance could be ignored in the following situations?

  **a** A marble dropped from an upstairs window.

  **b** A table tennis ball dropped from an upstairs window.

  **c** A marble dropped from an aircraft at 2000 metres altitude.

  **d** A child on a swing.

  **e** A person walking.

  **f** A person cycling.

2 Do you think it would be reasonable to disregard friction in the following situations?

  **a** Skiing in a straight line downhill.

  **b** Dragging a packing case across the floor.

  **c** Raising an object on a rope passing over a tree branch.

  **d** Raising an object on a rope passing over a pulley.

3 A loose tile slides down a sloping roof and falls to the ground. A question in a mechanics textbook describes the situation as follows. *A particle of mass 1.5 kg slides from rest down a smooth 30° slope of length 3 m, then falls to the ground 5 m below.*

  **a** What assumptions are being made in this description?

  **b** If the assumptions are wrong, how will the motion differ from that predicted?

4 The diagram shows an experiment done by a student to explore the impact between two rubber balls.

The balls are attached to two strings, which are tied to a fixed point $O$. Ball $A$ is allowed to swing down from a position level with $O$ and collides with ball $B$, which is hanging at rest below $O$. The student measures how far upwards ball $B$ rises after the impact.

  **a** What assumptions do you think the student should make? (Use conventional terms where appropriate.)

  **b** Discuss the effects on the results if the assumptions are not valid.

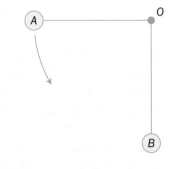

## Summary

Refer to

- When using mathematics to model a situation you
  - make simplifying assumptions
  - devise a mathematical model – usually equations – based on your assumptions
  - make predictions
  - obtain real data
  - compare predicted and actual results (allowing for errors)

  If the model is not good enough, change the assumptions and repeat the process.

  1.1

- Conventional terms are used to indicate the assumptions being made.
  - Objects of negligible size are described as 'small' or 'a particle'
  - Strings and rods may be 'light' (negligible mass) and 'inextensible' (do not stretch significantly)
  - Beams, objects and structures may be 'rigid' (do not bend or deform significantly)
  - A rigid beam or strut of negligible width is described as 'a rod'
  - Surfaces may be 'smooth' (friction can be ignored) or 'rough' (friction must be considered)
  - A flat sheet may be a 'lamina' (thickness is negligible).

  1.1

M1

---

### Links

Mathematical modelling is used in a wide variety of real-life situations. The National Airspace System (NAS) manages the air traffic flow at airports within the UK. Information about the speed and direction of incoming flights and conditions affecting their behaviour, such as weather patterns, are used to estimate arrival times several hours in advance. The NAS then uses radar signals to constantly update the model and ensure accurate information about the arrival and departure times of all flights.

# 2

## Motion in a straight line

This chapter will show you how to
- distinguish between a vector quantity and a scalar quantity
- calculate average speed and velocity
- draw and interpret graphs of displacement, velocity and acceleration against time
- use the constant acceleration equations, including the motion of a particle moving vertically under gravity.

## Before you start

### You should know how to:

1 Find the gradient of a straight line on a graph.

2 Find the area of rectangles, triangles and trapeziums.

3 Substitute values into an algebraic formula.

4 Solve a linear equation.

5 Solve a quadratic equation using factorisation or the quadratic formula.

6 Convert between metric units.

### Check in

1 Find the gradient of the line through the points $(1, 4)$ and $(3, 10)$.

2 A trapezium $ABCD$ has vertices $A(0, 0)$, $B(4, 9)$, $C(12, 9)$ and $D(15, 0)$. Find the area of $ABCD$.

3 In the equation $s = ut + \frac{1}{2}at^2$, find the value of $s$ when $u = 10$, $t = 4$ and $a = 6$.

4 Solve the equations
   a $5x - 3 = 3x + 8$
   b $\frac{1}{2}(x + 2) = \frac{1}{3}x + 2$

5 Solve the equations
   a $x^2 - 5x + 6 = 0$
   b $x^2 - 3x - 1 = 0$

6 Convert
   a $350$ g to kg
   b $108$ km h$^{-1}$ to m s$^{-1}$

M1

This line, marked in metres (m), shows an origin, *O*, and two points, *P* and *Q*.

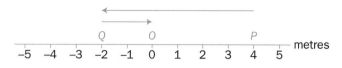

Suppose an object travels from *P* to *Q* and then to *O*. It moves from a position of 4 m at *P* to a position of –2 m at *Q* and then a position of 0 m at *O*.

The move from *P* to *Q* is a displacement of –6 m.
The move from *Q* to *O* is a displacement of 2 m.
The two moves give a combined displacement of –4 m.

Displacement is a vector quantity. A **vector quantity** has both magnitude (size) and direction.

Displacement is the change of position.

How far an object travels, ignoring direction, is called distance. This is **not** the same as displacement. The object in the diagram travels a total distance of 8 m in going from *P* to *Q* to *O*, but its resulting displacement from its starting point is –4 m.

Distance is a scalar quantity. A **scalar quantity** is one with magnitude only.

Adding the magnitude of the displacements gives the distance travelled.

---

**EXAMPLE 1**

A girl stands at the top of a 30 metre high cliff. She throws a ball vertically up so that it rises 20 metres then falls to the bottom of the cliff.

Find   **a**   the total displacement

      **b**   the total distance travelled.

The height of the girl's hand above the ground can be ignored.

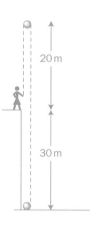

**a**   Take the starting point of the ball as the origin, and upwards as the positive direction:

Total displacement = 20 + (–50) = –30 m

The ball finishes 30 m below the origin.

**b**   Total distance = 20 + 50 = 70 m

## Velocity and speed

Suppose the same object is moving at a constant rate, taking 3 seconds to go from $P$ to $Q$ and 1 second from $Q$ to $O$.

– 6 m in 3 s

2 m in 1 s

$Q$    $O$       $P$    metres

–5   –4   –3   –2   –1   0   1   2   3   4   5

From $P$ to $Q$ its displacement changes by $-2$ m every second.
Its **velocity** is $-2$ metres per second ($\text{m s}^{-1}$).
From $Q$ to $O$ its displacement changes by $2$ m every second.
Its velocity is $2\ \text{m s}^{-1}$.

> **Velocity** is rate of change of displacement.

Velocity is a **vector** quantity. It indicates how fast the object travels and in which direction.

The **speed** of the object is $2\ \text{m s}^{-1}$ throughout the whole journey.

> **Speed** is the magnitude of the velocity.
> Speed is rate of change of distance.

Speed is a **scalar** quantity.

Consider a ball hitting a wall.

George rolls a ball at a wall 30 m away. It hits the wall after 3 s and rebounds towards George, who stops it 20 m from the wall after a further 5 s.

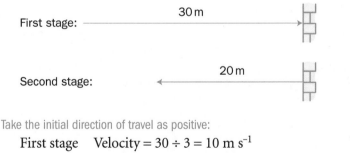

First stage:    30 m

Second stage:    20 m

Take the initial direction of travel as positive:

First stage    Velocity $= 30 \div 3 = 10\ \text{m s}^{-1}$
               Speed $= 30 \div 3 = 10\ \text{m s}^{-1}$

Second stage Velocity $= (-20) \div 5 = -4\ \text{m s}^{-1}$
               Speed $= 20 \div 5 = 4\ \text{m s}^{-1}$

You have assumed that the speed and velocity of the ball is constant during each stage.
If the motion of the ball varied during a stage, these results would give the **average speed** and **average velocity**, i.e. the constant speed and velocity needed to achieve the same distance and displacement in the same time.

Constant speed and velocity are also called **uniform** speed and velocity.

In reality objects rarely travel at a constant speed for any length of time, but the idea is a useful modelling assumption.

M1

$$\text{Average speed} = \frac{\text{total distance travelled}}{\text{total time taken}}$$

$$\text{Average velocity} = \frac{\text{total displacement}}{\text{total time taken}}$$

$$= \frac{\text{final position} - \text{initial position}}{\text{total time taken}}$$

You can find the average speed and velocity over several stages of a journey. Consider again the example of the ball hitting the wall:

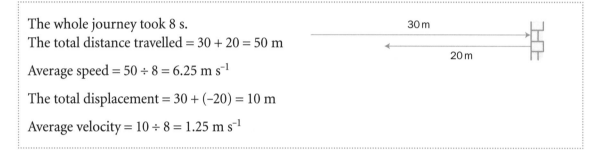

The whole journey took 8 s.
The total distance travelled = 30 + 20 = 50 m

Average speed = 50 ÷ 8 = 6.25 m s$^{-1}$

The total displacement = 30 + (−20) = 10 m

Average velocity = 10 ÷ 8 = 1.25 m s$^{-1}$

EXAMPLE 2

A car starts from a point $A$ and drives 300 m along a straight road to a point $B$. It then immediately reverses 120 m to a point $C$. It takes 15 s to go from $A$ to $B$ and 20 s from $B$ to $C$.

a Calculate the average speed
   i from $A$ to $B$   ii from $B$ to $C$   iii for the whole journey.

b Calculate the average velocity
   i from $A$ to $B$   ii from $B$ to $C$   iii for the whole journey.

c Express the result for **a iii** in km h$^{-1}$.

a i Average speed $= \dfrac{\text{distance travelled}}{\text{time taken}} = \dfrac{300}{15} = 20$ m s$^{-1}$

   ii Average speed $= \dfrac{120}{20} = 6$ m s$^{-1}$

   iii Average speed $= \dfrac{420}{35} = 12$ m s$^{-1}$

b Define the direction from $A$ to $B$ to be positive.

   i Average velocity $= \dfrac{\text{displacement}}{\text{time taken}} = \dfrac{300}{15} = 20$ m s$^{-1}$

   ii Average velocity $= \dfrac{-120}{20} = -6$ m s$^{-1}$

   iii The total displacement = 300 − 120 = 180 m, so

   Average velocity $= \dfrac{180}{35} = 5.14$ m s$^{-1}$

c 12 m s$^{-1}$ = 12 × 60 × 60 m h$^{-1}$ = 43 200 m h$^{-1}$ = 43.2 km h$^{-1}$

Even if the road had not been straight you could have modelled it as a straight line for the purpose of this calculation.

It is usual to model the car as a particle. You don't need to allow for the length of the car in your calculations.

The average speed of a journey of several stages cannot be found by calculating the mean of the average speeds for the individual stages. In this example 12 m s$^{-1}$ is not the mean of 20 m s$^{-1}$ and 6 m s$^{-1}$. The average speed for a journey must be found using the **total distance** and **total time**.

## Acceleration

**Acceleration** measures the amount by which velocity changes in each second. The unit of acceleration is metres per second per second, or metres per second squared ($m\ s^{-2}$).

> **Acceleration** is the rate of change of velocity.

Constant velocity corresponds to zero acceleration.

Acceleration is a **vector** quantity. It has a magnitude and a direction given by the change in velocity which can be positive or negative.

You can calculate the constant acceleration of an object:

> An object travelling at $1\ m\ s^{-1}$ undergoes constant acceleration, and 3 s later is travelling at $7\ m\ s^{-1}$.
>
> Velocity changes by $6\ m\ s^{-1}$ in 3 s.
> Acceleration $= 6 \div 3 = 2\ m\ s^{-2}$.
>
> An object travelling at $1\ m\ s^{-1}$ undergoes a constant acceleration of $-2\ m\ s^{-2}$ for 3 s.
>
> Velocity changes by $(-2) \times 3 = -6\ m\ s^{-1}$
> Final velocity $= 1 - 6 = -5\ m\ s^{-1}$.

Negative acceleration always decreases velocity, but the final speed may be greater than the initial speed.

The terms 'deceleration' or 'retardation' are sometimes used for negative acceleration. This is in situations where the velocity stays positive, and so the speed is reduced.

In the real world constant (uniform) acceleration rarely happens, but it is sometimes a useful modelling assumption.

If an object is not accelerating at a constant rate you can find the **average acceleration**.

> $$\text{Average acceleration} = \frac{\text{change in velocity}}{\text{time taken}}$$
>
> $$= \frac{\text{final velocity} - \text{initial velocity}}{\text{time taken}}$$

> The velocity of an object increases from $2\ m\ s^{-1}$ to $14\ m\ s^{-1}$ in 4 s.
>
> Average acceleration $= \dfrac{14 - 2}{4} = 3\ m\ s^{-2}$

Take care. This does not mean that the acceleration is constant at $3\ m\ s^{-2}$ over the 4 s.

M1

EXAMPLE 3

A hot air balloon is drifting at a constant velocity of 3 m s$^{-1}$.
A change in wind direction brings it to rest in 6 seconds.

**a** Calculate the average acceleration.

The wind continues to blow so that the balloon experiences
the same (constant) acceleration for a further 10 seconds.

**b** Find the velocity and speed of the balloon at the end of this time.

**a** The balloon's velocity changes from 3 m s$^{-1}$ to 0 m s$^{-1}$ in 6 s.

$$\text{Average acceleration} = \frac{\text{final velocity} - \text{initial velocity}}{\text{time taken}} = \frac{0 - 3}{6} = -0.5 \text{ m s}^{-2}$$

**b** In each second the velocity changes by $-0.5$ m s$^{-1}$.
The balloon starts the 10 seconds at rest.

Final velocity $= 0 + 10 \times (-0.5) = -5$ m s$^{-1}$, final speed $= 5$ m s$^{-1}$

## Exercise 2.1

1 A sentry on guard duty marches 50 m east from his sentry box.
He then goes 90 m in a westerly direction before finally returning
to the sentry box. Taking the sentry box as the origin and
east as the positive direction, find

    **a**  his position at the end of each stage

    **b**  the displacement he undergoes in each of the three stages

    **c**  the total distance for the whole journey

    **d**  the total displacement for the whole journey.

2 A car travels 150 km between 10.30 am and 12.20 pm. Find its average speed

    **a**  in km h$^{-1}$               **b**  in m s$^{-1}$.

3 In a charity walk, a group walked for 2 hours, covering a distance
of 6 km. They then stopped for lunch, which took an hour, and
afterwards walked the final 12 km in 3 hours. Find, in km h$^{-1}$, their
average speed for the whole journey.

4 A cyclist travels 12 km at a speed of 15 km h$^{-1}$ and then continues for 36
minutes at a speed of $26\frac{2}{3}$ km h$^{-1}$. Find her average speed

    **a**  in km h$^{-1}$               **b**  in m s$^{-1}$.

5 A man jogs 8 km at a constant speed of 6 km h$^{-1}$, then cycles for 2 hours
at a constant speed of 16 km h$^{-1}$. Find, in km h$^{-1}$, his average speed

    **a**  for the first 12 km       **b**  for the whole journey.

M1

6 A car travels 2 km at an average speed of 20 m s$^{-1}$, then 2 km at an average speed of 25 m s$^{-1}$.

   a  Find the average speed for the whole journey.

   b  Explain why, having finished the first stage at 20 m s$^{-1}$, the car could not achieve an average speed of 40 m s$^{-1}$ for the whole journey.

7 A canoe race involves paddling 1800 m upstream and then 1200 m downstream to the finish. A competitor takes 10 minutes for the upstream section and returns at 5 m s$^{-1}$.

   Find   a  the average speed for the upstream section
          b  the time for the downstream section
          c  the average speed for the whole race
          d  the average velocity for the whole race.

8 A spaceship is travelling at 15 m s$^{-1}$. The forward thrusters are activated, giving an acceleration of $-2$ m s$^{-2}$.

   a  How long will it take for the spaceship to come to rest?

   b  If the thrusters fire for a total of 20 s, find the final velocity and speed of the spaceship.

9 A ball bearing is rolled up a slope. Its initial speed is 4 m s$^{-1}$. After 8 seconds it is rolling down the slope at 6 m s$^{-1}$. Find its average acceleration during this time.

10 A lift, parked on the third floor, is called by someone on the seventh floor, who then gets in and travels to the ground floor. Each floor is 3.5 m and the whole process takes 50 s.

   a  Find the overall average speed of the lift.

   b  Find the overall average velocity of the lift.

   c  The lift passed the third floor at 3 m s$^{-1}$ on the way down. If it came to rest at the ground floor 12 seconds later, find its average acceleration for that part of the journey.

Take the upward direction as positive.

11 Sharon drove a distance of 18 km from her home to the motorway before joining it for the rest of her journey. Her average speed for the first part of the journey was 15 m s$^{-1}$ and her overall average speed was $22\frac{2}{9}$ m s$^{-1}$. The whole journey took 72 minutes. Find

   a  the total distance she travelled

   b  her average speed for the second part of the journey.

12 Amanda's car has a maximum speed of $2V$ m s$^{-1}$. She wishes to complete a certain journey at an average speed of $V$ m s$^{-1}$. What is the minimum average speed at which she could complete the first half of the journey and still meet her target for the whole journey?

M1

You can illustrate graphically the motion of an object in several ways. The simplest option is to plot its displacement (from the origin) against time.

EXAMPLE 1

A tiger pacing along the front of her enclosure moves in one direction at 2 m s⁻¹ for 6 seconds, then turns and goes the other way at 3 m s⁻¹ for 5 seconds. Represent this on a displacement–time graph.

The symbol *s* is commonly used for displacement. The graph can be called an *s*–*t* graph (or sometimes a *t*, *s* graph).

Take the first direction as positive and the tiger's starting point as the origin.

First-stage displacement = 6 × 2 = 12 m

Second-stage displacement = 5 × (–3) = –15 m

The tiger moves from a position of 0 m to a position of 12 m and then to a position of –3 m.

Use this information to draw the graph.
Plot time on the horizontal axis and displacement on the vertical axis.

The velocities here are uniform (constant), giving straight lines on the graph.
Non-uniform velocity would give a curved graph.

You could also draw a distance–time graph. Both sections would then slope upwards, as the distance travelled is always increasing.

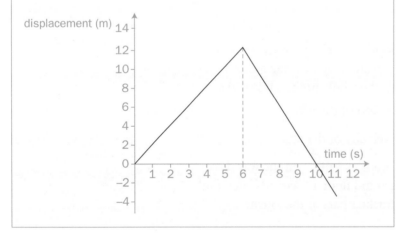

### Velocity from a displacement–time graph

You know that $\dfrac{\text{change of displacement}}{\text{time taken}} = \text{velocity}$,

but on the displacement–time graph

$$\frac{\text{change of displacement}}{\text{change of time}} = \text{gradient}.$$

The **gradient** of a graph gives the **rate of change** of the quantity. Velocity is rate of change of displacement.

It follows that

> The gradient of a displacement–time graph gives the velocity of the object.

If the graph is not a straight line, the gradient of the tangent to the curve at any point is the velocity of the object at that instant.

e.g. The gradients of the two sections of the graph in Example 1 are 2 and –3. The velocities were 2 m s$^{-1}$ and –3 m s$^{-1}$.

The gradient of a distance–time graph is the speed.

**EXAMPLE 2**

Find  **a**  the average speed

**b**  the average velocity

of the tiger in Example 1.

**a**  Average speed = $\dfrac{\text{total distance travelled}}{\text{time taken}}$

The distance travelled is

$$(2 \times 6) + (3 \times 5) = 27 \text{ m}$$
$$\text{so average speed} = \frac{27}{11}$$
$$= 2.45 \text{ m s}^{-1}$$

**b**

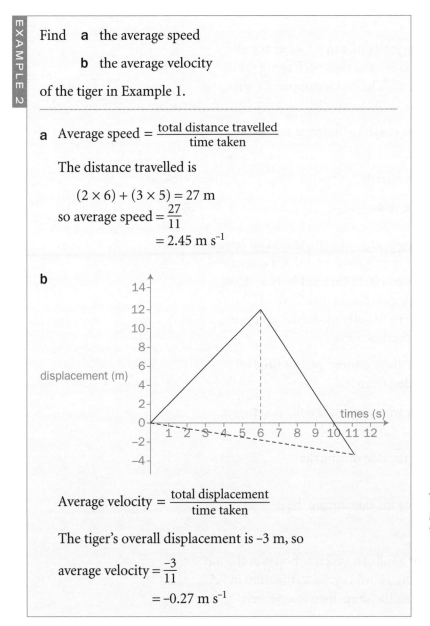

Average velocity = $\dfrac{\text{total displacement}}{\text{time taken}}$

The tiger's overall displacement is –3 m, so

$$\text{average velocity} = \frac{-3}{11}$$
$$= -0.27 \text{ m s}^{-1}$$

The average velocity is the gradient of the dotted line on the graph.

M1

## Exercise 2.2

1 The graph shows the displacement (in km) of a cyclist
   from a town, plotted against time (in hours).

   **a** Describe the journey, giving the cyclist's velocity
   during each stage.

   **b** State three assumptions that have been made
   in drawing this graph.

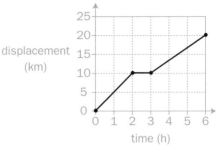

2 A ball is rolled at a constant velocity of 4 m s$^{-1}$. After travelling
   6 m it strikes a wall at right angles and rolls back along the
   same line at a constant 2.5 m s$^{-1}$. The ball is stopped 3 s after
   hitting the wall.

   **a** Draw a displacement–time graph to illustrate the ball's
   motion.

   **b** Calculate the ball's average speed.

   **c** Calculate the ball's average velocity.

3 A woman, walking a dog along a straight path, stops and releases
   the dog at a point $A$. She then continues forward at a constant
   speed of 1.4 m s$^{-1}$. The dog runs 100 m forward in 10 s, stops
   and sniffs for 10 s, then runs forward a further 50 m in 20 s.
   It then spots another dog 100 m the other side of $A$ and runs
   back to join it at a constant speed of 5 m s$^{-1}$.

   **a** Draw a graph to show the displacement against time of
   both the woman and the dog from $A$.

   **b** From your graph estimate where and when the dog passes
   the woman.

   **c** Find the average speed of the dog during the
   period described.

   **d** Find the average velocity of the dog during the
   period described.

4 Harold lives at the bottom of a hill. The village shop is at the top
   of the hill, a distance of 1.2 km. Harold cycles to the shop in 7.5
   minutes, spends 3.5 minutes in the shop, then free-wheels
   home at a speed of 4 m s$^{-1}$.

   **a** Draw a displacement–time graph to illustrate Harold's
   shopping trip, stating any assumptions that you make.

   **b** Calculate Harold's average speed for the whole journey.

   **c** What is Harold's average velocity for the whole journey?

5   A cyclist starts from town *A* at 11 am to ride to town *B* 60 km away. He completes the journey in three stages of 20 km, each taking an hour, with 15 minute breaks between stages. A second cyclist starts from *B* at the same time as the first cyclist leaves *A*, and travels to *A* non-stop at 16 km h$^{-1}$.

   a   On the same axes, draw graphs showing the displacements against time of the two cyclists from *A*.

   b   At what time, and where, do the two cyclists pass each other?

6   At 10.00 am Adam joined a motorway at Junction 3 and travelled without stopping at 80 km h$^{-1}$ to Junction 36, 380 km north.
   At the same time Marlon joined the motorway at Junction 5, 30 km north of Junction 3. He drove north at 90 km h$^{-1}$. After 150 km he took a 50 minute break, then continued his journey north, leaving the motorway at Junction 36 at 2.10 pm.

   a   On the same axes draw distance–time graphs for Adam's and Marlon's journeys.

   b   Between what times was Adam ahead of Marlon?

   c   What was Marlon's average speed after he took his break?

7   A particle starts from rest and travels with non-uniform velocity. Its displacements at intervals of 1 s were noted, as shown in the table.

| Time (s) | 0 | 1 | 2 | 3 | 4 | 5 | 6 | 7 | 8 | 9 | 10 |
|---|---|---|---|---|---|---|---|---|---|---|---|
| Displacement (m) | 0 | 2.9 | 6.9 | 12.0 | 18.3 | 25.7 | 34.3 | 44.0 | 54.9 | 66.9 | 80.0 |

   a   Draw a displacement–time graph to represent this data.

   b   Use your graph to estimate the speed of the particle

      i   after 1 s
      ii  after 7 s.

   c   Find the average speed of the particle over the first 7 s.

You can also draw a graph of velocity against time.
Plot time on the horizontal axis and velocity on the vertical axis.

EXAMPLE 1

An object starts from rest and accelerates uniformly to a velocity of 6 m s⁻¹ in 15 seconds. It then undergoes a constant negative acceleration so that after a further 20 seconds its velocity is –4 m s⁻¹. Draw a graph of velocity against time.

The acceleration is uniform during each stage, so the graph consists of two straight lines.

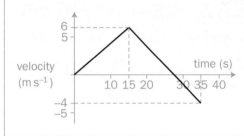

The symbol $v$ is commonly used for velocity. The graph can be called a $v$–$t$ graph (or sometimes a $t$, $v$ graph).

Straight lines on the graph correspond to uniform (constant) acceleration.
Non-uniform acceleration would give a curved graph.

## Acceleration from a velocity–time graph
You know that

$$\frac{\text{change of velocity}}{\text{time taken}} = \text{acceleration},$$

but on the velocity–time graph

$$\frac{\text{change of velocity}}{\text{change of time}} = \text{gradient}.$$

It follows that

> The gradient of a velocity–time graph gives the acceleration of the object.

The **gradient** of a graph gives the **rate of change** of the quantity. Acceleration is the rate of change of velocity.

This can be seen in Example 1. During the first stage the velocity increased by 6 m s⁻¹ in 15 s. This is an acceleration of 0.4 m s⁻², and the gradient of the line on the graph is 0.4.
During the second stage the velocity changed by –10 m s⁻¹ in 20 s. This is an acceleration of –0.5 m s⁻², and the gradient of the line on the graph is –0.5.

If the graph is not a straight line, the gradient of the tangent of the curve at any point is the acceleration of the object at that instant.

## Speed–time graphs

You could plot speed rather than velocity.

EXAMPLE 2

Draw a speed–time graph for the situation described in
Example 1.

As speed is a scalar quantity, its value is always
positive, leading to the graph shown.

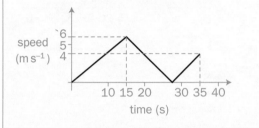

## Displacement from a velocity–time graph

You can relate displacement to a velocity–time graph, as shown
by this example.

A cyclist travelling at a constant velocity of 8 km h$^{-1}$ for
3 h achieves a displacement of 24 km.
The velocity–time graph is shown.

The shaded area between the graph and the time axis is
$3 \times 8 = 24$.
**This area corresponds to the displacement.**

If the cyclist had been travelling in the negative direction, the
velocity would have been –8 km h$^{-1}$, as shown, and the
displacement would have been –24 km.

The shaded area is now $3 \times (-8) = -24$.

The area again corresponds to the displacement.

This relationship also holds true for non-uniform velocity.

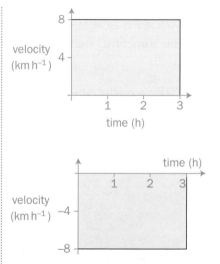

You may know from your study of
calculus that you obtain negative
results when finding areas below
the horizontal axis by integration.

The area between a velocity–time graph and the time axis
gives the displacement of the object.

The area under a speed–time
graph corresponds to the
distance travelled.

M1

For the situation described in Example 1, calculate

**a** the overall displacement and

**b** the total distance travelled by the object.

**a** When time = 15 s, velocity = 6 m s$^{-1}$.
Acceleration from then on is –0.5 m s$^{-2}$,
so $v = 0$ after a further 12 s.
Hence velocity = 0 m s$^{-1}$ when time = 27 s.

The triangular region above the axis has base 27
and height 6.
Its area = 81, so displacement = 81 m.

The triangular region below the axis has base 8
and height –4.
Its area = –16, so displacement = –16 m.

Overall displacement = 81 m + (–16) m = 65 m.

**b** The total distance travelled = 81 m + 16 m = 97 m.

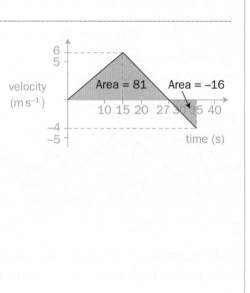

## Acceleration–time graphs

It is sometimes useful to draw an acceleration–time graph.

The acceleration–time graph corresponding to the
velocity–time graph in Example 1 would look like this:

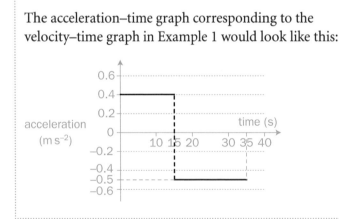

This graph makes the modelling
assumption that acceleration
changes instantaneously from one
value to another.

The area between the
acceleration–time graph and the
time axis gives the change of
velocity. You should check that
this is true for this graph.

You can summarise the features of the three types of motion
graphs in a table.

| | Displacement–time graph | Velocity–time graph | Acceleration–time graph |
|---|---|---|---|
| **Gradient gives:** | Velocity | Acceleration | – |
| **Area gives:** | – | Displacement | Change of velocity |

## Exercise 2.3

1 The displacement–time graph shows the progress of a hiker out on a day's trek.

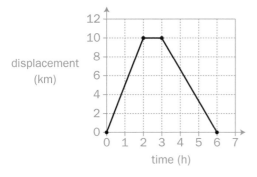

**a** Describe the motion during each stage of the journey.

**b** Draw the corresponding velocity–time graph.

2 A car, travelling at 15 m s$^{-1}$, accelerates uniformly to a velocity of 45 m s$^{-1}$ in 12 s.

**a** Sketch the velocity–time graph.

**b** Calculate the car's acceleration.

**c** Calculate the distance the car travels while accelerating.

3 A lorry, travelling at 36 m s$^{-1}$, is brought uniformly to rest with acceleration –1.2 m s$^{-2}$.

**a** Sketch the velocity–time graph.

**b** Calculate the distance travelled by the lorry before it stops.

4 A lift starts from rest, accelerates upwards for 2 seconds at 1.5 m s$^{-2}$, travels for 3 seconds at constant speed, and then decelerates to rest in 1.2 seconds.

**a** Sketch the velocity–time graph, stating any assumptions you have made.

**b** Sketch the corresponding acceleration–time graph.

**c** Calculate the displacement of the lift between stops.

M1

5 The graph shows the acceleration of an object during a period of 7 seconds. At the start of the period the velocity of the object is 1 m s⁻¹.

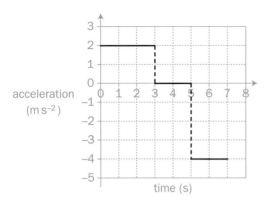

a Sketch the velocity–time graph.

b Calculate the overall displacement of the object.

c Calculate the distance travelled by the object.

6 A dragonfly, at rest on a bullrush, decides to fly to a second bullrush 18 m away. It accelerates uniformly to a speed of 5 m s⁻¹, then immediately decelerates uniformly to rest on the second bullrush.  Sketch the velocity–time graph and find how long the journey takes.

7 A boat starts from rest at a point $A$ and accelerates uniformly at a rate of 0.5 m s⁻² for 12 seconds. The boat then decelerates uniformly to rest in 15 seconds. The boat then accelerates as it reverses back towards $A$. It reaches a speed of 4 m s⁻¹ in 20 seconds and then continues to reverse at that speed.

a Sketch the velocity–time graph.

b What was the boat's greatest forward displacement from $A$?

c What was the total time between the boat's leaving $A$ and its return to $A$?

8 A car starts from rest at a point $A$ and drives up a slope. It accelerates to a speed of 18 m s$^{-1}$ in 6 s and maintains this speed for 4 s. The gears are then disengaged and the car coasts to rest with acceleration $-2$ m s$^{-2}$. Unfortunately, the driver forgets to put the handbrake on, so the car then rolls back down the slope with acceleration $-1$ m s$^{-2}$.

  a  Sketch a velocity–time graph of the motion.

  b  Calculate the acceleration on the first stage.

  c  Find how far the car was from $A$ when it came instantaneously to rest.

  d  When will it return to $A$?

9 A car, $P$, is at rest at a point $O$ when a second car, $Q$, passes at a constant speed of 20 m s$^{-1}$. At the moment that $Q$ passes, $P$ moves off in pursuit. It accelerates uniformly at 2 m s$^{-2}$ until it reaches a speed of 30 m s$^{-1}$, then continues at this speed.

  a  Sketch the velocity–time graph for both cars on the same axes.

  b  How far is $P$ behind $Q$ at the moment when it reaches full speed?

  c  How much longer does it take for $P$ to draw level with $Q$?

10 Chulchit drives to work along a straight road of length 8 km. His car accelerates and decelerates at 2.5 m s$^{-2}$, and his preferred cruising speed is 90 km h$^{-1}$.

  a  Assuming he has a clear run with no hold-ups, sketch a velocity–time graph of his journey and hence calculate his journey time.

  b  Chulchit hears on the radio that there are 2 km of road works, with a speed limit of 36 km h$^{-1}$, somewhere along the route. Investigate the effect of the positioning of these road works on his best journey time, and find the maximum and minimum values of this.

M1

Problems concerning the motion of objects involve some or all of the following quantities.

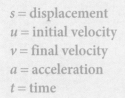

$s$ = displacement
$u$ = initial velocity
$v$ = final velocity
$a$ = acceleration
$t$ = time

If you can assume that acceleration is constant, these quantities are connected together by five simple formulae.

You can derive two of these formulae directly from the velocity–time graph.

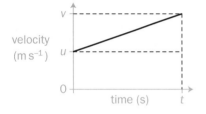

Acceleration = gradient of velocity–time graph.

This gives   $a = \dfrac{v - u}{t}$

$\Rightarrow$       $v = u + at$           [1]

Displacement = area under velocity–time graph.

Use the formula for the area of a trapezium:

$s = \dfrac{1}{2}(u + v)t$           [2]

You can combine equations [1] and [2] to give three more formulae.

Substitute for $v$ from [1] into [2]:

$s = \dfrac{1}{2}(u + u + at)t$

$\Rightarrow$   $s = ut + \dfrac{1}{2}at^2$           [3]

M1

From [1] you have $u = v - at$

Substitute this into [2]:

$$s = \frac{1}{2}(v - at + v)t$$

$$\Rightarrow \quad s = vt - \frac{1}{2}at^2 \qquad\qquad [4]$$

From [1] you have $t = \dfrac{v - u}{a}$

Substitute this into [2]:

$$s = \frac{(u + v)(v - u)}{2a}$$

$$\Rightarrow \quad v^2 = u^2 + 2as \qquad\qquad [5]$$

You will use some formulae more frequently than others.
The complete list, in order of usefulness, is:

$$v = u + at$$

$$s = ut + \frac{1}{2}at^2$$

$$v^2 = u^2 + 2as$$

$$s = \frac{1}{2}(u + v)t$$

$$s = vt - \frac{1}{2}at^2$$

Each formula links four of the five variables. You need to learn these formulae.

When acceleration is **not uniform** these formulae are **not valid** and should **not be used**.

When solving problems, be systematic:

o   write down the variable whose value you are trying to find
o   write down three variables whose values you know
o   choose the formula which links these four variables.

EXAMPLE 1

A car, travelling at a speed of 15 m s$^{-1}$, suddenly accelerates at 3 m s$^{-2}$. What is its speed after 5 seconds?

You need to find $v$.
You know the values $u = 15$ m s$^{-1}$, $a = 3$ m s$^{-2}$ and $t = 5$ s.
The formula containing these variables is $v = u + at$.

Substitute the known values:

$$v = 15 + 3 \times 5 = 30$$

So the car's speed after 5 seconds is 30 m s$^{-1}$.

EXAMPLE 2

A hot air balloon is drifting at a constant velocity of 3 m s$^{-1}$. A change in wind causes it to undergo an acceleration of $-0.5$ m s$^{-2}$ for a period of 16 s.
Calculate

**a** the displacement of the balloon

**b** the distance travelled by the balloon during this period.

For problems involving vector quantities you should choose one direction to be positive and be consistent.

**a** You need to find $s$.
You know the values $u = 3$ m s$^{-1}$, $a = -0.5$ m s$^{-2}$ and $t = 16$ s.

The formula containing these variables is $s = ut + \frac{1}{2}at^2$

Substitute the known values:

$$s = 3 \times 16 + \frac{1}{2} \times (-0.5) \times 16^2 = -16$$

so the overall displacement is $-16$ m.

**b** To find the distance travelled you need to know how far forward the balloon went before it began to move backwards.
This is the point at which the balloon came instantaneously to rest.

You need to find $s$.
You know the values $u = 3$ m s$^{-1}$, $v = 0$ m s$^{-1}$ and $a = -0.5$ m s$^{-2}$.
The formula containing these variables is $v^2 = u^2 + 2as$

Substitute the known values: $0 = 3^2 + 2 \times (-0.5)s$
$$\Rightarrow \quad s = 9$$

So the balloon moved 9 m forward.
It then drifted backwards 25 m to achieve an overall displacement of $-16$ m.

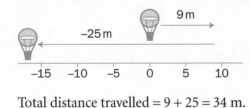

Total distance travelled $= 9 + 25 = 34$ m.

EXAMPLE 3

A car, travelling at a speed of 10 m s$^{-1}$, accelerates at 4 m s$^{-2}$ until its speed has increased to 18 m s$^{-1}$. How far does it travel while accelerating?

You need to find $s$.
You know the values $u = 10$ m s$^{-1}$, $v = 18$ m s$^{-1}$ and $a = 4$ m s$^{-2}$.
The formula containing these variables is $\quad v^2 = u^2 + 2as$

Substitute the known values: $\quad 324 = 100 + 8s$
$$\Rightarrow \quad s = 28$$

So the distance travelled during the acceleration is 28 m.

EXAMPLE 4

A lorry is travelling at 30 m s$^{-1}$. The driver applies the brakes and the lorry comes to rest after 5 seconds. How far did it travel during this time?

You need to find $s$.
You know the values $u = 30$ m s$^{-1}$, $v = 0$ m s$^{-1}$ and $t = 5$ s.

The formula containing these variables is $\quad s = \frac{1}{2}(u + v)t$

Substitute the known values: $\quad s = \frac{1}{2}(30 + 0) \times 5$
$$= 75$$

So the distance travelled is 75 m.

EXAMPLE 5

A particle, accelerating at 10 m s$^{-2}$, strikes a wall at 42 m s$^{-1}$. How far did it travel in the final second of its motion?

You need to find $s$.
You know the values $v = 42$ m s$^{-1}$, $a = 10$ m s$^{-2}$ and $t = 1$ s.

The formula containing these variables is $\quad s = vt - \frac{1}{2}at^2$

Substitute the known values: $\quad s = 42 \times 1 - \frac{1}{2} \times 10 \times 1^2$
$$= 37$$

So the distance it travelled in the final second is 37 m.

M1

Some problems involve more than one object.

EXAMPLE 6

A car, $P$, is accelerating at $2$ m s$^{-2}$. At the point where its velocity is $10$ m s$^{-1}$ it is overtaken by another car, $Q$, travelling at $16$ m s$^{-1}$ and accelerating at $1$ m s$^{-2}$. How long is it before car $P$ catches up with car $Q$?

You need to find the value of $t$ for which the displacements of the two cars are equal.

For car $P$ you have $u = 10$ m s$^{-1}$ and $a = 2$ m s$^{-2}$.
At time $t$ its displacement is

$$s_P = 10 \times t + \frac{1}{2} \times 2 \times t^2 = 10t + t^2$$

For car $Q$ you have $u = 16$ m s$^{-1}$ and $a = 1$ m s$^{-2}$.
At time $t$ its displacement is

$$s_Q = 16 \times t + \frac{1}{2} \times 1 \times t^2 = 16t + \frac{1}{2}t^2$$

The cars are level when $s_P = s_Q$

$$\Rightarrow \quad 10t + t^2 = 16t + \frac{1}{2}t^2$$

$$t^2 - 12t = 0$$

$$t(t - 12) = 0 \quad \Rightarrow \quad t = 0 \text{ or } t = 12$$

$t = 0$ is the time when $Q$ passed $P$, so $P$ catches up with $Q$ 12 seconds later.

## Exercise 2.4

1  A lorry starts from rest and accelerates uniformly at a rate of $3$ m s$^{-2}$ for $30$ s.

   a  How far does it travel?

   b  How fast is it travelling at the end of the period?

2  A stone is dropped from rest at the top of a tower. It takes 5 seconds to reach the ground, by which time it is travelling at $50$ m s$^{-1}$.

   a  What is its acceleration?

   b  How high is the tower?

3  A body starts from rest with uniform acceleration and in $10$ s moves a distance of $150$ m.

   a  What is its acceleration?

   b  How fast is it moving at the end of this period?

4  A stunt motorcyclist has 50 m in which to accelerate from rest to the 90 km h$^{-1}$ needed at the ramp. How long does this 'run up' take?

5  A train leaves station $A$ from rest with constant acceleration 0.2 m s$^{-2}$. It reaches maximum speed after 2 minutes, maintains this speed for 4 minutes, then slows down to stop at station $B$ with acceleration −1.5 m s$^{-2}$. Calculate the distance $AB$.

6  A train accelerates uniformly from rest for 1 minute, at the end of which time its velocity is 30 km h$^{-1}$. It maintains this speed until it is 500 m from the next station. It then decelerates uniformly and stops at the station. Calculate the train's acceleration during the first and last phases of this journey.

7  A car crosses a speed hump with a velocity of 4 m s$^{-1}$. It then accelerates at a rate of 2.5 m s$^{-2}$ to a speed of 9 m s$^{-1}$ when the driver applies the brakes, causing an acceleration of −3 m s$^{-2}$, reducing the speed of the car to 4 m s$^{-1}$ to cross the next hump.

   a  How far apart are the humps?

   b  How long does the car take to travel from one hump to the next?

   c  The question implies that the car is being modelled as a particle. In what way does this assumption affect your results?

8  A moon landing craft is 1 km above the lunar surface and descending at a speed of 80 m s$^{-1}$. The rockets are then fired, giving it an upward acceleration $a$ m s$^{-2}$. Find the value of $a$ if the craft is to make a perfect soft landing.

9  A car starts from rest at the bottom of a slope. It accelerates up the slope for 8 seconds at 1.5 m s$^{-2}$, then disengages the engine and coasts. If its acceleration is now −1 m s$^{-2}$, find the time between the car's leaving the bottom of the slope and returning to it.

10  A lift ascends from rest with an acceleration of 0.5 m s$^{-2}$ before slowing with an acceleration of −0.75 m s$^{-2}$ for the next stop. If the total journey time is 10 s, what is the distance between the two stops?

M1

11  A boat is travelling at a speed of 4 m s$^{-1}$. Its propeller is then put into reverse, giving it an acceleration of –0.4 m s$^{-2}$ for a period of 25 seconds.

    **a**  Find the overall displacement of the boat during this period.

    **b**  Find the distance travelled by the boat during this period.

12  An object travels 10 m during one second and 15 m during the next second.

    **a**  Find the acceleration of the object, assuming it to be constant.

    **b**  How fast is the object going at the end of the two seconds?

13  An object moving with constant acceleration travels 10 m in 2 seconds. The next 10 m takes it 4 seconds.

    **a**  Find the acceleration of the object.

    **b**  For how much more time will the object travel before coming to rest?

    **c**  How much further will the object travel before coming to rest?

14  Theresa is handing the baton on to Magda in a relay race. Theresa is running at a constant speed and when she is 4.5 m away Magda starts running with acceleration 1 m s$^{-2}$. Theresa continues at a constant speed and just manages to catch up and hand over the baton. How fast was Theresa running?

15  Trains $A$ and $B$ are travelling in the same direction on two parallel straight tracks. At the moment that $A$ passes $B$, $A$ is travelling at a constant 20 m s$^{-1}$ and $B$ is travelling at 5 m s$^{-1}$ and accelerating at 0.5 m s$^{-2}$.

    **a**  How much time passes before the trains are again level?

    **b**  How far do they travel in this time?

16  Clare is driving along a road in her car, with Henry following 40 m behind in his car. They are both travelling at a speed of 25 m s$^{-1}$. Clare spots a problem ahead and brakes to a halt with an acceleration of –5 m s$^{-2}$. Henry takes 0.2 seconds to react, then brakes, but his brakes are poor and only give an acceleration of –4 m s$^{-2}$. Investigate what happens.

**17** $P$ is a point 10 m from the bottom of a slope. A ball is rolled up the slope from $P$ with an initial velocity of 8 m s$^{-1}$. It undergoes a constant acceleration of $-4$ m s$^{-2}$.

    **a i** How long does it take to reach the bottom of the slope?

      **ii** How far up the slope does it travel?

    **b** One second after the first ball is rolled, a second ball is rolled up from the bottom of the slope. It has an initial velocity of 14 m s$^{-1}$. It also has an acceleration of $-4$ m s$^{-2}$. Find when and where the two balls meet.

**18** A particle travelling in a straight line with acceleration $a$ passes a point $A$ whilst moving with velocity $u$. When it reaches a point $B$ it has velocity $v$, and at this point its acceleration changes to $-a$. Show that when it again passes through $A$ its speed is $\sqrt{2v^2 - u^2}$.

**19** A car, $C$, travelling at constant speed $u$, passes a stationary police car, $P$, which immediately sets off in pursuit with acceleration $a$. Find, in terms of $u$ and $a$, the greatest distance separating the two cars during the subsequent chase.

**20** Two cars are travelling at a constant speed $v$, with one car a distance $d$ ahead of the other. As each car passes a marker post it accelerates with acceleration $a$ to a new constant speed $V$. Show that the new distance $D$ between the cars is $D = \dfrac{Vd}{v}$.

**21** Two trains, $A$ and $B$, are travelling on parallel tracks and in opposite directions. They start simultaneously from rest at stations a distance $d$ m apart and head towards each other, with constant accelerations 0.4 m s$^{-2}$ and 0.2 m s$^{-2}$ respectively. After 50 s the fronts of the trains are level. Calculate

    **a** the speed of each train when they meet

    **b** the value of $d$.

M1

## Free fall under gravity

You can assume that an object moving vertically under the effect of gravity has constant acceleration provided

- the object is so small and the speed is so low that air resistance can be neglected
- the distance it moves is small relative to the size of the Earth, so that the Earth's gravitational field can be assumed to be constant.

The acceleration due to gravity is denoted by $g$.

Near the Earth's surface the acceleration due to gravity is approximately

$$g \approx 9.8 \text{ m s}^{-2}$$

Sometimes a question will use the approximation $g = 10 \text{ m s}^{-2}$, but it will be clearly stated when this is the case.

**EXAMPLE 1**

A stone is dropped from the top of a 20 m high tower.

**a** How long does it take to reach the ground?

**b** With what velocity does it hit the ground?

Take the origin as the top of the tower and take downwards as the positive direction.
You know $u = 0 \text{ m s}^{-1}$, $a = g = 9.8 \text{ m s}^{-2}$ and $s = 20$ m.

**a** Use $s = ut + \frac{1}{2}at^2$ to find $t$:

$$20 = \frac{1}{2} \times 9.8 \times t^2$$

$$\Rightarrow \quad t = 2.02 \quad \text{or} \quad -2.02$$

You can obviously ignore the negative value, so $t = 2.02$ s

**b** Use $v^2 = u^2 + 2as$ to find $v$:

$$v^2 = 2 \times 9.8 \times 20$$

$$= 392$$

$$\Rightarrow \quad v = 19.8 \quad \text{or} \quad -19.8$$

Again you can ignore the negative value, so $v = 19.8 \text{ m s}^{-1}$.

Having found the value of $t$ in part **a**, you could have found $v$ using $v = u + at$.

EXAMPLE 2

A ball is thrown vertically upwards from ground level at 20 m s$^{-1}$. A boy, leaning out of a window 8 m above the point of projection, catches the ball on its way down.

**a** What is the time of flight of the ball?

**b** How fast is it travelling when the boy catches it?

Take the origin as ground level and take upwards as the positive direction.

You know $u = 20$ m s$^{-1}$, $a = g = -9.8$ m s$^{-2}$.

**a** When the ball is caught, $s = 8$ m.

Use $s = ut + \frac{1}{2}at^2$ to find $t$:

$$8 = 20t + \frac{1}{2} \times (-9.8) \times t^2$$

$$\Rightarrow \quad 4.9t^2 - 20t + 8 = 0$$

$$\Rightarrow \quad t = 3.63 \quad \text{or} \quad 0.450$$

The ball is level with the boy at $t = 0.450$ going up, and at $t = 3.63$ going down.

So, the time of flight is 3.63 s.

**b** Use $v = u + at$ to find $v$:

$$v = 20 + (-9.8) \times 3.63$$
$$= -15.6$$

So, the ball is travelling at 15.6 m s$^{-1}$ when it is caught. The negative sign shows that it is travelling downwards.

> You could have used $v^2 = u^2 + 2as$, giving $v = 15.6$ (going up) or $-15.6$ (going down).

MI

## Exercise 2.5

1 A stone is dropped from the top of a 50 m high cliff.

  **a** How long does it take to reach the beach below?

  **b** With what velocity does it hit the beach?

2 A ball is thrown vertically upwards with speed 15 m s$^{-1}$.

  **a** Find the greatest height it reaches.

  **b** How long does it take to reach this maximum height?

  **c** The ball returns to its starting position. What is the whole time of flight?

3   A ball thrown vertically upwards rises 20 m before descending again.

   a   What was its initial speed?

   b   What is the whole time of flight?

   c   With what speed is the ball travelling when it arrives back at its starting point?

4   A stone, thrown vertically upwards at 5 m s$^{-1}$ from the edge of a 60 m high cliff, falls to the beach below.

   a   With what speed does it hit the beach?

   b   What is the time of flight?

5   A stone is thrown vertically upwards with speed 10 m s$^{-1}$. One second later another stone is thrown vertically upwards from the same point and with the same speed.

   a   How high are the stones when they meet?

   b   How long after the first stone is thrown do the stones meet?

6   A boy drops a stone from the top of a building of height $h$. Simultaneously his friend throws another stone vertically upwards from the ground below, with a speed of 30 m s$^{-1}$. The stones meet 1.5 seconds later. Find $h$.

7   A body falls from rest from the top of a tower. During the last second of its motion it falls $\frac{7}{16}$ of the whole distance.

   a   Show that the time of descent is independent of the value of $g$.

   b   Find the height of the tower in terms of $g$.

8   A rocket ascends vertically from rest at ground level with acceleration 10 m s$^{-2}$. Calculate

   a   the height of the rocket after 4 s

   b   the speed of the rocket after 4 s.

   After the rocket has been moving for 4 s a component breaks off and falls to the ground.

   c   Calculate the further time which elapses before the component hits the ground.

9 A ball is projected vertically upwards from ground level at a speed of 30 m s$^{-1}$.

Calculate

 a the height to which the ball rises

 b the time for which the ball is in the air

 c the length of time for which the ball is over 30 m above the ground.

10 A stone falls past a window of height 2.5 m in 0.5 s. Taking $g = 10$ m s$^{-2}$, find the height from which the stone fell.

11 An object is thrown vertically downwards with speed $V$. During the sixth second of its motion it travels a distance $h$. Find $V$ in terms of $h$ and $g$.

12 An object is projected vertically upwards with a velocity $u$ m s$^{-1}$. $T$ seconds later another object is projected vertically upwards from the same point and with the same speed. Find, in terms of $u$, $T$ and $g$, the further time which elapses before the objects collide.

13 a A particle is projected vertically upwards with initial speed $u$ m s$^{-1}$. After $T$ s it reaches its greatest height, $H$ m. Find, in terms of $u$ and $g$, an expression for

   i $T$   ii $H$

 b At a certain time the particle is travelling upwards through a point $A$, at a height of 19.6 m, and 2 s later it is at a point $B$, at a height of 29.4 m.

   i Show that it is on its way down through $B$ at that time.
   ii Find the value of $u$.

14 A particle $A$ is thrown vertically upwards from the bottom of a tower. At the same instant a second particle, $B$, is dropped from the top of the tower. Given that when the particles collide they are travelling at the same speed, find the ratio between the distances they have travelled.

1 A train travels from rest at station *A*. It moves for 1 minute
with a constant acceleration of 0.5 m s$^{-2}$, continues at uniform
speed for 3 minutes, and then slows to rest at station *B* in a further 30 s.

  **a** Sketch a velocity–time graph to show the train's journey.

  **b** Calculate the maximum speed of the train during the journey.

  **c** Calculate the distance between the two stations.

  **d** Calculate the train's average speed.

2 A runner moves from rest at a constant acceleration of 0.6 m s$^{-2}$
for 10 s, then continues at a uniform velocity. A cyclist starts from
rest at the same point and time and accelerates uniformly for 8 s
before immediately slowing uniformly to rest after a further 22 s.
At the moment that the cyclist stops they have both travelled
the same distance.

  **a** Sketch a velocity–time graph for the runner.

  **b** Sketch a velocity–time graph for the cyclist.

  **c** Calculate the distance travelled by the cyclist.

  **d** Calculate the greatest speed reached by the cyclist.

3

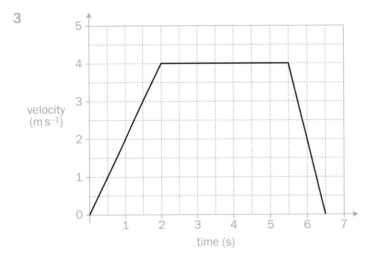

The graph shows the velocity of a lift travelling upwards between
two floors of a building.

  **a** State any assumptions which have been made in drawing this graph.

  **b** Calculate the acceleration of the lift during each phase of the motion.

  **c** Calculate the distance between the two floors.

**4**

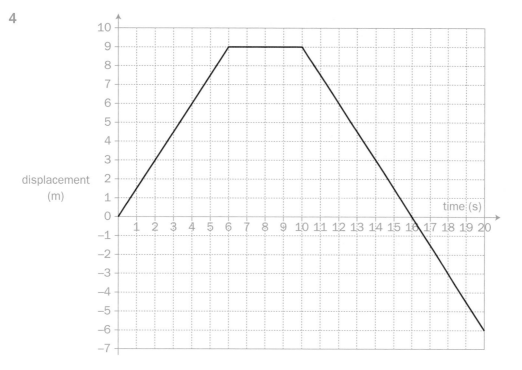

The graph shows the motion of a cat along the top of a straight fence.

**a** Describe the motion of the cat during this period of 20 s.

**b** Calculate the velocity of the cat during each phase of the motion.

**c** Calculate the average speed of the cat during this period of 20 s.

**d** Calculate the average velocity of the cat during this period of 20 s.

**5** A car moves in a straight line for 6 s at a constant acceleration of
2.5 m s⁻². The brakes are then applied, bringing it to rest in 4 s.
It is immediately put into reverse, accelerating uniformly so that it
passes its original starting point travelling at a speed of 6 m s⁻¹.

**a** Sketch a velocity–time graph for the car's journey.

**b** Calculate the greatest positive displacement of the car
from its starting point.

**c** Calculate the total time for the motion described.

**d** Calculate the car's acceleration when travelling in reverse.

**e** Sketch an acceleration–time graph for the motion described.

6 A particle is sliding over a rough plane. Its initial speed is 12 m s$^{-1}$ and its acceleration is $-2$ m s$^{-2}$.

    **a** Calculate how long it takes to travel 20 m.

    **b** Calculate how far it travels before coming to rest.

7 A tobogganer puts her feet down to try to avoid hitting a snowdrift that she sees 30 m ahead. She achieves a retardation of 0.2 m s$^{-2}$, but this is not enough to stop her before she reaches the snowdrift. She runs into the snowdrift with a speed of 2 m s$^{-1}$.

    **a** For how long does she have her feet down?

    **b** How fast was she travelling when she first put her feet down?

8 A competitor makes a dive from a high springboard into a diving pool. She leaves the springboard vertically with a speed of 4 m s$^{-1}$ upwards. When she leaves the springboard, she is 5 m above the surface of the pool. The diver is modelled as a particle moving vertically under gravity alone and it is assumed that she does not hit the springboard as she descends. Find

    **a** her speed when she reaches the surface of the pool

    **b** the time taken to reach the surface of the pool.

    **c** State two physical factors which have been ignored in the model.

                                                   [(c) Edexcel Limited 2003]

9 A small ball is projected vertically upwards from a point $A$. The greatest height reached by the ball is 30 m above $A$.

Calculate

    **a** the speed of projection

    **b** the time between the instant that the ball is projected and the instant it returns to $A$.

10 A car moves with constant acceleration along a straight horizontal road. The car passes the point $A$ with speed 5 m s$^{-1}$ and 4 s later it passes the point $B$, where $AB = 50$ m.

a  Find the acceleration of the car.

When the car passes the point $C$, it has speed 30 m s$^{-1}$.

b  Find the distance $AC$.                                    [(c) Edexcel Limited 2002]

11 A ball is dropped from the top of a cliff and hits the beach 4 seconds later.

a  Assuming that there is no air resistance, calculate the height of the cliff.

b  If air resistance in fact has a significant effect, would the actual height of the cliff be greater or less than that calculated in part a? Explain your answer.

12 A car starts from rest at a point $S$ on a straight racetrack. The car moves with constant acceleration for 20 s, reaching a speed of 25 m s$^{-1}$. The car then travels at a constant speed of 25 m s$^{-1}$ for 120 s. Finally it moves with constant deceleration, coming to rest at a point $F$.

a  Sketch a speed–time graph to illustrate the motion of the car.

The distance between $S$ and $F$ is 4 km.

b  Calculate the total time the car takes to travel from $S$ to $F$.

A motorcycle starts at $S$, 10 s after the car has left $S$. The motorcycle moves with constant acceleration from rest and passes the car at a point $P$ which is 1.5 km from $S$. When the motorcycle passes the car, the motorcycle is still accelerating and the car is moving at a constant speed. Calculate

c  the time the motorcycle takes to travel from $S$ to $P$

d  the speed of the motorcycle at $P$.                        [(c) Edexcel Limited 2002]

M1

13 The stopping distance of a car consists of thinking distance plus braking distance. Harold's car can brake with a deceleration of 5 m s$^{-2}$. There is a 1 s delay between Harold's decision to stop and his foot pressing the brake.

   **a** Find an expression for Harold's stopping distance when he is driving at a speed of $u$ m s$^{-1}$.

   **b** If Harold sees an obstacle in the road 60 m ahead, for what range of values of $u$ will he be able to stop in time?

14 Two cars, travelling in the same direction, are level as they pass a point $A$. At this stage the first car is travelling at 12 m s$^{-1}$, while the second car has a speed of 6 m s$^{-1}$. They both accelerate uniformly, the first car at 0.4 m s$^{-2}$ and the second at 0.6 m s$^{-2}$. The cars are again level when they pass a point $B$.

Calculate

   **a** the time they take to travel from $A$ to $B$

   **b** the distance $AB$

   **c** the speeds of the two cars at $B$.

15 Cars $A$ and $B$ are approaching the end of a race. $A$ is 1.5 km from the finish, is travelling at a speed of 30 m s$^{-1}$ and is accelerating uniformly at 0.7 m s$^{-2}$. $B$ is 210 m behind $A$, is travelling at a speed of 40 m s$^{-1}$ and is accelerating at 0.5 m s$^{-2}$.

   **a** Show that $B$ overtakes $A$ 285 m before the finish.

   **b** Calculate the difference in time between the arrivals of the two cars at the finish.

16 Arnold and Bruno are running for a bus, which pulls away from the bus-stop when they are 10 m short of reaching it. The bus has a uniform acceleration of 0.8 m s$^{-2}$.

   **a** Arnold runs at a constant speed of 5 m s$^{-1}$. Find how far the bus has moved before he catches it.

   **b** Bruno also runs at a constant speed but is slower than Arnold and only just catches the bus (that is, the bus has the same speed as Bruno at the moment that he draws level). Find Bruno's running speed.

17 A parachutist drops from a helicopter $H$ and falls vertically from rest
towards the ground. Her parachute opens 2 s after she leaves $H$ and her
speed then reduces to 4 m s$^{-1}$. For the first 2 s her motion is modelled as
that of a particle falling freely under gravity. For the next 5 s the model
is motion with constant deceleration, so that her speed is 4 m s$^{-1}$ at the
end of this period. For the rest of the time before she reaches the ground,
the model is motion with constant speed of 4 m s$^{-1}$.

a Sketch a speed–time graph to illustrate her motion from $H$
to the ground.

b Find her speed when the parachute opens.

A safety rule states that the helicopter must be high enough
to allow for the parachute to open and for the speed of a
parachutist to reduce to 4 m s$^{-1}$ before reaching the ground.

Using the assumptions made in the above model,

c find the minimum height of $H$ for which the woman can
make a drop without breaking this safety rule.

Given that $H$ is 125 m above the ground when the woman
starts her drop

d find the total time taken for her to reach the ground.

e State one way in which the model could be refined to
make it more realistic.                              [(c) Edexcel Limited 2001]

18

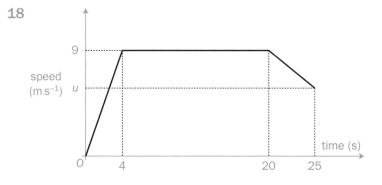

A sprinter runs a race of 200 m. Her total time for running the
race is 25 s. The diagram is a sketch of the speed–time graph
for the motion of the sprinter. She starts from rest and accelerates
uniformly to a speed of 9 m s$^{-1}$ in 4 s. The speed of 9 m s$^{-1}$ is
maintained for 16 s and she then decelerates uniformly to a
speed of $u$ m s$^{-1}$ at the end of the race. Calculate

a the distance covered by the sprinter in the first 20 s of the race

b the value of $u$

c the deceleration of the sprinter in the last 5 s of the race.       [(c) Edexcel Limited 2005]

## Summary

Refer to

- Quantities used to measure motion are vectors or scalars.
  - Displacement, velocity and acceleration are vectors.
  - Distance and speed are scalars.
- You can find average values for speed, velocity and acceleration.  2.1
- You can draw graphs of displacement and distance against time.
  - The gradient of a displacement–time graph gives velocity.
  - The gradient of a distance–time graph gives speed.  2.2
- You can draw graphs of velocity, speed and acceleration against time.
  - The gradient of a velocity–time graph gives acceleration.
  - The area under a velocity–time graph gives displacement.
  - The area under an acceleration–time graph gives change of velocity.  2.3
- When acceleration can be assumed uniform (constant), the displacement ($s$), initial velocity ($u$), final velocity ($v$), acceleration ($a$) and time ($t$) are connected by these five equations:

$$v = u + at$$

$$s = ut + \frac{1}{2}at^2$$

$$v^2 = u^2 + 2as$$   You need to memorise these.

$$s = \frac{1}{2}(u + v)t$$

$$s = vt - \frac{1}{2}at^2$$   2.4

- Objects moving vertically and freely under gravity have uniform acceleration of $g \approx 9.8 \text{ m s}^{-2}$.  2.5

### Links

Railway operators need to calculate train accelerations and decelerations in order to plan their timetables, to position signals allowing for adequate stopping distances and, most importantly, to ensure the safe and efficient operation of railways.

Trains must be powerful enough to accelerate to the line speed limit within a short space of time as well as having reliable brakes to enable them to come to a standstill at a station or a signal as and when is needed.

# 3

## Vectors

This chapter will show you how to

- manipulate vectors in two dimensions
- express vectors in component form
- find the magnitude and direction of a vector expressed in component form
- solve problems involving the position vectors of objects moving in two dimensions with constant velocity
- solve problems involving the velocity vectors of objects moving in two dimensions with constant acceleration.

## Before you start

### You should know how to:

1   Solve problems using Pythagoras' theorem and the trigonometry of right-angled triangles.

2   Solve quadratic equations using factorisation and the quadratic formula.

3   Understand and use bearings to solve problems involving angles.

### Check in

1   A triangle $ABC$ has angle $C = 90°$.

   **a**  Find $AC$, given that $AB = 14$ cm and $BC = 9$ cm.

   **b**  Find $AB$, given that $AC = 8$ cm and angle $A = 34°$.

   **c**  Find angle $B$, given that $AC = 6$ cm and $BC = 9$ cm.

2   Solve

   **a**  $x^2 - 3x - 40 = 0$

   **b**  $3x^2 + 2 = 7x$

   **c**  $4x^2 + 5x - 2 = 0$

3   A boat sets off from a harbour at point $O$. It travels 7 km at a bearing of 140° to a point $A$. The boat then travels to point $B$ which is 3.5 km on a bearing of 260° from point $A$. Find

   **a**  the bearing the boat should take to travel back to the harbour in a straight line from $B$

   **b**  the distance of the homeward journey described in part **a**.

In Chapter 2 you met the terms vector and scalar.

> A **vector quantity** has both magnitude (size) and direction.
>
> A **scalar quantity** has magnitude only.

e.g. Displacement, velocity and acceleration are vector quantities.

e.g Distance and speed are scalar quantities.

The easiest vector to illustrate is a displacement, or translation, by a given distance in a given direction. The mathematical techniques you learn when working with displacements can also be applied to velocities and other vector quantities.

You can represent a displacement by a directed line segment.

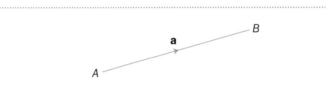

This arrow represents a translation from $A$ to $B$.
The translation from $A$ to $B$ is written as $\overrightarrow{AB}$, whereas the distance fron $A$ to $B$ is written as $AB$.
Alternatively you can label the vector with a single letter in bold type, such as **a**. This would be handwritten as $\underline{a}$.
You write the magnitude of the vector $\overrightarrow{AB}$ as $|\overrightarrow{AB}|$ or $AB$.
You write the magnitude of the vector **a** as $|a|$ or $a$.

A **unit vector** is a vector with a magnitude of 1.
The unit vector in the direction of a vector **a** is usually labelled $\hat{\mathbf{a}}$.

You call this vector 'a hat'.

> Vectors are equal if they have the same magnitude and the same direction.

In this parallelogram
$\overrightarrow{AB} = \overrightarrow{DC}$
$\overrightarrow{AD} = \overrightarrow{BC}$

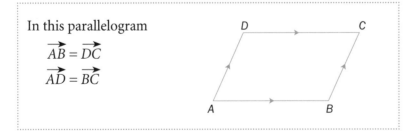

M1

Combining the translations $\overrightarrow{AB}$ and $\overrightarrow{BC}$ has the same effect as the single translation $\overrightarrow{AC}$.

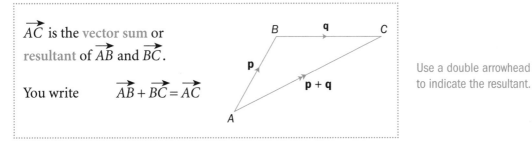

$\overrightarrow{AC}$ is the vector sum or resultant of $\overrightarrow{AB}$ and $\overrightarrow{BC}$.

You write $\qquad \overrightarrow{AB} + \overrightarrow{BC} = \overrightarrow{AC}$

Use a double arrowhead to indicate the resultant.

The zero vector, **0**, has zero magnitude. Its direction is undefined.

The vector **–a** has the same magnitude as **a** but the opposite direction.

The resultant of $\overrightarrow{AB}$ and $\overrightarrow{BA}$ is **0**.
$$\overrightarrow{AB} + \overrightarrow{BA} = 0$$
You can write $\qquad \overrightarrow{BA} = -\overrightarrow{AB}$

If you multiply a vector by a scalar, you change its magnitude but not its direction.

$k\mathbf{a}$ is a vector parallel to **a** and with magnitude $k|\mathbf{a}|$.

e.g. $2\mathbf{a} = \mathbf{a} + \mathbf{a}$, which is the translation **a** applied twice.
The result gives a translation twice as far in the same direction.

The resultant of two or more vectors is unchanged by the order in which you apply or bracket them.

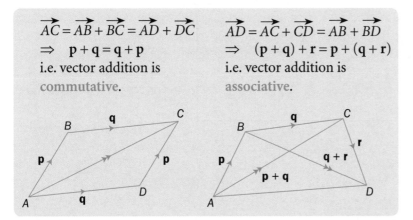

$\overrightarrow{AC} = \overrightarrow{AB} + \overrightarrow{BC} = \overrightarrow{AD} + \overrightarrow{DC}$
$\Rightarrow \quad \mathbf{p} + \mathbf{q} = \mathbf{q} + \mathbf{p}$
i.e. vector addition is commutative.

$\overrightarrow{AD} = \overrightarrow{AC} + \overrightarrow{CD} = \overrightarrow{AB} + \overrightarrow{BD}$
$\Rightarrow \quad (\mathbf{p} + \mathbf{q}) + \mathbf{r} = \mathbf{p} + (\mathbf{q} + \mathbf{r})$
i.e. vector addition is associative.

Subtracting a vector is the same as adding its negative:

$$\mathbf{p} - \mathbf{q} = \mathbf{p} + (-\mathbf{q})$$

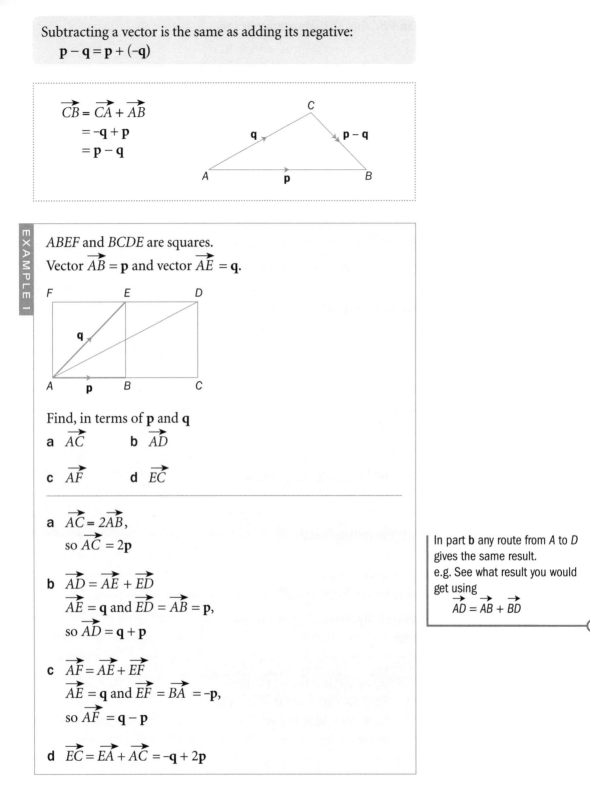

$$\overrightarrow{CB} = \overrightarrow{CA} + \overrightarrow{AB}$$
$$= -\mathbf{q} + \mathbf{p}$$
$$= \mathbf{p} - \mathbf{q}$$

**EXAMPLE 1**

*ABEF* and *BCDE* are squares.

Vector $\overrightarrow{AB} = \mathbf{p}$ and vector $\overrightarrow{AE} = \mathbf{q}$.

Find, in terms of **p** and **q**

**a** $\overrightarrow{AC}$      **b** $\overrightarrow{AD}$

**c** $\overrightarrow{AF}$      **d** $\overrightarrow{EC}$

---

**a** $\overrightarrow{AC} = 2\overrightarrow{AB}$,

  so $\overrightarrow{AC} = 2\mathbf{p}$

**b** $\overrightarrow{AD} = \overrightarrow{AE} + \overrightarrow{ED}$

  $\overrightarrow{AE} = \mathbf{q}$ and $\overrightarrow{ED} = \overrightarrow{AB} = \mathbf{p}$,

  so $\overrightarrow{AD} = \mathbf{q} + \mathbf{p}$

**c** $\overrightarrow{AF} = \overrightarrow{AE} + \overrightarrow{EF}$

  $\overrightarrow{AE} = \mathbf{q}$ and $\overrightarrow{EF} = \overrightarrow{BA} = -\mathbf{p}$,

  so $\overrightarrow{AF} = \mathbf{q} - \mathbf{p}$

**d** $\overrightarrow{EC} = \overrightarrow{EA} + \overrightarrow{AC} = -\mathbf{q} + 2\mathbf{p}$

In part **b** any route from *A* to *D* gives the same result.

e.g. See what result you would get using

$$\overrightarrow{AD} = \overrightarrow{AB} + \overrightarrow{BD}$$

M1

You can use vector methods to prove many standard geometrical theorems.

EXAMPLE 2

$ABCD$ is a parallelogram. $E$ is the mid-point of $AC$.
Vector $\overrightarrow{AB} = \mathbf{p}$ and vector $\overrightarrow{AD} = \mathbf{q}$.

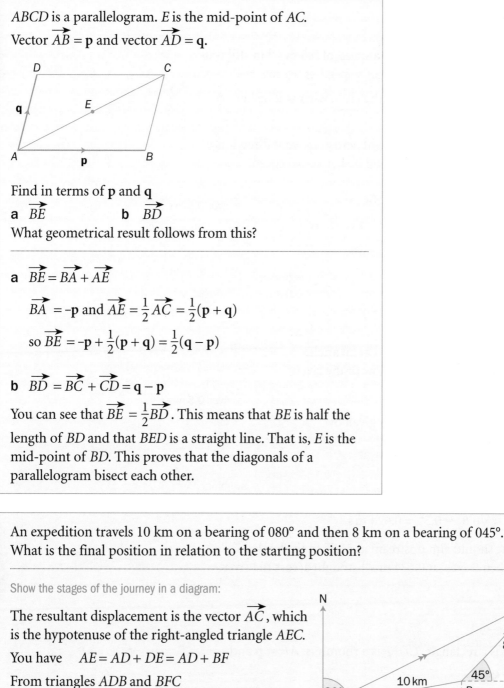

Find in terms of $\mathbf{p}$ and $\mathbf{q}$
**a** $\overrightarrow{BE}$          **b** $\overrightarrow{BD}$
What geometrical result follows from this?

**a** $\overrightarrow{BE} = \overrightarrow{BA} + \overrightarrow{AE}$

$\overrightarrow{BA} = -\mathbf{p}$ and $\overrightarrow{AE} = \frac{1}{2}\overrightarrow{AC} = \frac{1}{2}(\mathbf{p} + \mathbf{q})$

so $\overrightarrow{BE} = -\mathbf{p} + \frac{1}{2}(\mathbf{p} + \mathbf{q}) = \frac{1}{2}(\mathbf{q} - \mathbf{p})$

**b** $\overrightarrow{BD} = \overrightarrow{BC} + \overrightarrow{CD} = \mathbf{q} - \mathbf{p}$

You can see that $\overrightarrow{BE} = \frac{1}{2}\overrightarrow{BD}$. This means that $BE$ is half the
length of $BD$ and that $BED$ is a straight line. That is, $E$ is the
mid-point of $BD$. This proves that the diagonals of a
parallelogram bisect each other.

EXAMPLE 3

An expedition travels 10 km on a bearing of 080° and then 8 km on a bearing of 045°.
What is the final position in relation to the starting position?

Show the stages of the journey in a diagram:

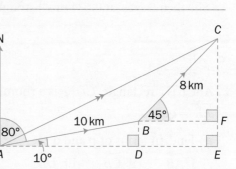

The resultant displacement is the vector $\overrightarrow{AC}$, which
is the hypotenuse of the right-angled triangle $AEC$.

You have    $AE = AD + DE = AD + BF$

From triangles $ADB$ and $BFC$
       $AD = 10\cos 10°$ and $BF = 8\cos 45°$
      $\Rightarrow AE = 10\cos 10° + 8\cos 45° = 15.5$ km

Similarly, $EC = EF + FC = 10\sin 10° + 8\sin 45° = 7.39$ km

So, $AC = \sqrt{15.5^2 + 7.39^2} = 17.2$ km

The bearing of $\overrightarrow{AC} = 090° - C\hat{A}E$, where $\tan C\hat{A}E = \frac{7.39}{15.5}$

The bearing of $\overrightarrow{AC} = 090° - 25.5° = 064.5°$

So $\overrightarrow{AC}$ is a displacement of 17.2 km on a bearing of 064.5°.

You could solve this problem by
applying the sine and cosine rules
to the triangle $ABC$. You will study
these rules in module C2.

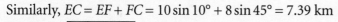

You solve problems involving other vector quantities, such as velocities, in the same way. You draw vector diagrams in which the vectors appear as displacements.

EXAMPLE 4

A swimmer, who can swim at a speed of 0.8 m s$^{-1}$ in still water, wishes to cross a river flowing at a speed of 0.5 m s$^{-1}$.

**a** If she aims straight across the river, what will be her actual velocity?

**b** If she wishes to travel straight across, in what direction should she aim and what will be her actual speed?

---

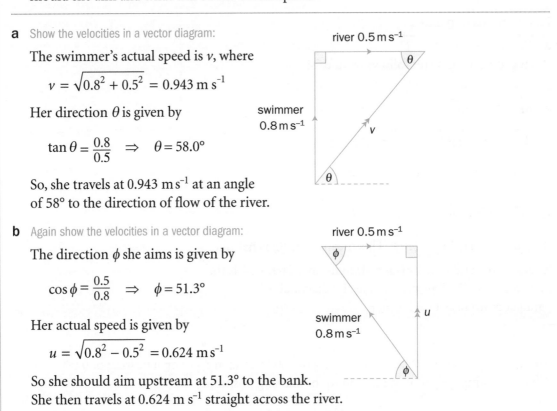

**a** Show the velocities in a vector diagram:

The swimmer's actual speed is $v$, where

$$v = \sqrt{0.8^2 + 0.5^2} = 0.943 \text{ m s}^{-1}$$

Her direction $\theta$ is given by

$$\tan \theta = \frac{0.8}{0.5} \implies \theta = 58.0°$$

So, she travels at 0.943 m s$^{-1}$ at an angle of 58° to the direction of flow of the river.

**b** Again show the velocities in a vector diagram:

The direction $\phi$ she aims is given by

$$\cos \phi = \frac{0.5}{0.8} \implies \phi = 51.3°$$

Her actual speed is given by

$$u = \sqrt{0.8^2 - 0.5^2} = 0.624 \text{ m s}^{-1}$$

So she should aim upstream at 51.3° to the bank. She then travels at 0.624 m s$^{-1}$ straight across the river.

## Exercise 3.1

**1** *ABCE* is a rectangle. *CDEF* is a rhombus. $\overrightarrow{AF} = \mathbf{p}$ and $\overrightarrow{EB} = \mathbf{q}$.

**a** Find, in terms of $\mathbf{p}$ and $\mathbf{q}$

i $\overrightarrow{AB}$    ii $\overrightarrow{CB}$    iii $\overrightarrow{DB}$

**b** Show that $\overrightarrow{EB} + \overrightarrow{CA} = 2\overrightarrow{DF}$.

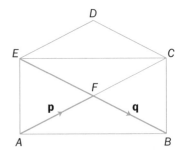

**2** The diagram shows a regular hexagon *ABCDEF*.
$\overrightarrow{AB} = \mathbf{p}$ and $\overrightarrow{BC} = \mathbf{q}$.

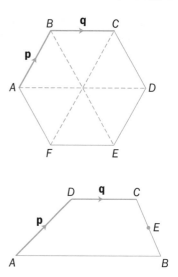

Find, in terms of **p** and **q**

**a** $\overrightarrow{AD}$     **b** $\overrightarrow{AC}$     **c** $\overrightarrow{CE}$

**d** $\overrightarrow{BE}$     **e** $\overrightarrow{EA}$     **f** $\overrightarrow{CA}$

**3** The diagram shows a trapezium *ABCD*, with *AB* parallel to
*DC* and *AB* twice as long as *DC*.

*E* is the mid-point of *BC*. $\overrightarrow{AD} = \mathbf{p}$ and $\overrightarrow{DC} = \mathbf{q}$

Find, in terms of **p** and **q**

**a** $\overrightarrow{AB}$       **b** $\overrightarrow{AC}$       **c** $\overrightarrow{CD}$

**d** $\overrightarrow{DB}$       **e** $\overrightarrow{AE}$       **f** $\overrightarrow{ED}$

**4** *ABC* is a triangle. *E* and *F* are the mid-points of *AB* and *AC*
respectively.
$\overrightarrow{AB} = \mathbf{p}$ and $\overrightarrow{AC} = \mathbf{q}$.

**a** Find, in terms of **p** and **q**
  **i** $\overrightarrow{BC}$     **ii** $\overrightarrow{EF}$

**b** What can you deduce about the lines *BC* and *EF*?

**5** *ABCD* is a quadrilateral. *E* and *F* are the mid-points of the
diagonals *AC* and *BD* respectively.

**a** Show that $\overrightarrow{AB} + \overrightarrow{AD} = 2\overrightarrow{AF}$ and $\overrightarrow{CB} + \overrightarrow{CD} = 2\overrightarrow{CF}$.

**b** Hence show that $\overrightarrow{AB} + \overrightarrow{AD} + \overrightarrow{CB} + \overrightarrow{CD} = 4\overrightarrow{EF}$.

**6** In each of the following cases, find the magnitude and direction
of the resultant of the two vectors.

**a** A displacement of 3.5 km on a bearing of 050° and a displacement
of 5.4 km on a bearing of 128°.

**b** A displacement of 26 km on a bearing of 175° and a displacement
of 18 km on a bearing of 294°.

**c** Velocities of 15 km h$^{-1}$ due north and 23 km h$^{-1}$ on a
bearing of 253°.

**d** Forces of 355 N on a bearing of 320° and 270 N on a
bearing of 025°.

        N stands for newtons, the unit of
        force, which you will meet in 4.1.

M1

7 Two ships, *A* and *B*, leave port *O* simultaneously. Ship *A* travels north at 16 km h$^{-1}$, and ship *B* travels due east at 13 km h$^{-1}$.

**a** The vector $\overrightarrow{AB}$ represents the displacement of B from *A*. Express $\overrightarrow{AB}$ in terms of $\overrightarrow{OA}$ and $\overrightarrow{OB}$.

**b** Find the magnitude and direction of $\overrightarrow{AB}$ after

**i** 1 hour     **ii** 3 hours     **iii** *t* hours.

**c** The ships' radios have a range of 120 km. For how long will the ships remain in contact?

**d** For how long would they have remained in contact if *B* had travelled north-east?

8 A boat, which can travel at 5 m s$^{-1}$ in still water, is crossing a river 200 m wide. The rate of flow of the river is 2 m s$^{-1}$ (which is assumed to be the same at all points of the river). Points *A* and *B* are on the banks and directly opposite each other across the river.

**a** If the boat leaves *A* and steers towards *B*, at what speed will it travel and how far downstream will it reach the other bank?

**b** If the boat needs to travel directly from *A* to *B*, in what direction should it be steered and at what speed will it travel?

9 An aircraft has a speed in still air of 400 km h$^{-1}$. A wind is blowing from the south at 80 km h$^{-1}$.

**a** The pilot steers the aircraft due east. Find the actual speed and direction of travel.

**b** The pilot wishes the plane to travel due east. Find the direction in which the aircraft should be steered and the speed at which it will travel.

10 A ship is being steered due east. A current is flowing from north to south so that the ship actually travels at 12 km h$^{-1}$ on a bearing of 120°. Find the speed of the current and the speed of the ship in still water.

11 A boat *P*, capable of 6 m s$^{-1}$ in still water, travels from a point *A* around a course *ABC* which is an equilateral triangle with sides 500 m long. A uniform current of 4 m s$^{-1}$ flows in the direction $\overrightarrow{AB}$. An identical boat *Q* starts from *A* at the same time and travels the other way (*ACB*) around the course. Which boat gets back to *A* first, and what is the margin of victory (in seconds)?

12  A river is $D$ m wide and flows at $u$ m s$^{-1}$. A man can swim at $v$ m s$^{-1}$ in still water, where $v > u$. Find the ratio of the shortest time it would take him to swim across the river and back, and the time it would take him to swim $D$ m upstream and back.

13  Harold and Imelda can both swim at 3 m s$^{-1}$. They are standing at point $A$ on the bank of a river, which is 90 m wide and flowing at 2 m s$^{-1}$. They wish to get across to the point $B$, immediately opposite them on the other bank.

Harold enters the water at $A$ and swims so that he travels straight towards $B$.

**a**  In what direction will he need to swim?

**b**  How long will it take him to reach $B$?

Imelda aims straight towards $B$, which means she will be carried downstream and land at a point $C$. She then runs along the bank to $B$.

**c**  Find the distance $BC$.

**d**  If they both arrive at $B$ at the same time, how fast did Imelda run?

14  $OABC$ is a tetrahedron. $L, M$ and $N$ are the mid-points of $OA, OB$ and $OC$ respectively. $P, Q$ and $R$ are the mid-points of $BC, AC$ and $AB$ respectively. $\overrightarrow{OA} = \mathbf{a}, \overrightarrow{OB} = \mathbf{b}$ and $\overrightarrow{OC} = \mathbf{c}$. Use vector methods to show that the lines $PL, QM$ and $RN$ bisect each other.

A unit vector is a vector with magnitude 1.

> The unit vector in the $x$-direction is called **i**.
> The unit vector in the $y$-direction is called **j**.

Rather than specify a vector using its magnitude and direction, you can state it in component form in terms of **i** and **j**.

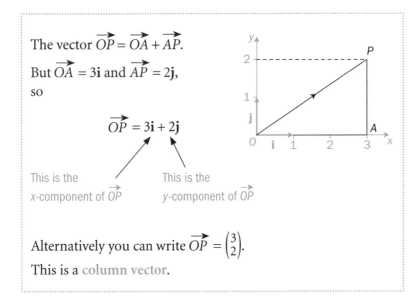

The vector $\overrightarrow{OP} = \overrightarrow{OA} + \overrightarrow{AP}$.

But $\overrightarrow{OA} = 3\mathbf{i}$ and $\overrightarrow{AP} = 2\mathbf{j}$,

so

$$\overrightarrow{OP} = 3\mathbf{i} + 2\mathbf{j}$$

This is the
$x$-component of $\overrightarrow{OP}$

This is the
$y$-component of $\overrightarrow{OP}$

Alternatively you can write $\overrightarrow{OP} = \begin{pmatrix} 3 \\ 2 \end{pmatrix}$.

This is a column vector.

When you are given a vector as a magnitude and direction, you can convert it to component form.
This is called resolving the vector into components.

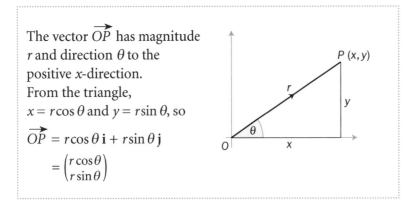

The vector $\overrightarrow{OP}$ has magnitude $r$ and direction $\theta$ to the positive $x$-direction.
From the triangle,
$x = r\cos\theta$ and $y = r\sin\theta$, so

$$\overrightarrow{OP} = r\cos\theta\,\mathbf{i} + r\sin\theta\,\mathbf{j}$$
$$= \begin{pmatrix} r\cos\theta \\ r\sin\theta \end{pmatrix}$$

M1

When you are given a vector in component form, you can calculate its magnitude and direction.

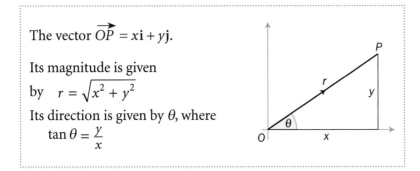

The vector $\overrightarrow{OP} = x\mathbf{i} + y\mathbf{j}$.

Its magnitude is given

by $r = \sqrt{x^2 + y^2}$

Its direction is given by $\theta$, where

$\tan \theta = \dfrac{y}{x}$

EXAMPLE 1

For the diagrams shown, express the vectors $\overrightarrow{OP}$, $\overrightarrow{OQ}$ and $\overrightarrow{OR}$ in component form.

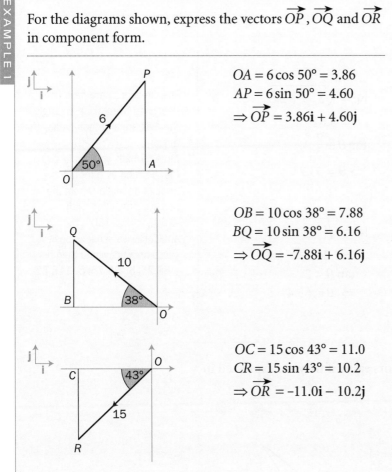

$OA = 6\cos 50° = 3.86$
$AP = 6\sin 50° = 4.60$
$\Rightarrow \overrightarrow{OP} = 3.86\mathbf{i} + 4.60\mathbf{j}$

$OB = 10\cos 38° = 7.88$
$BQ = 10\sin 38° = 6.16$
$\Rightarrow \overrightarrow{OQ} = -7.88\mathbf{i} + 6.16\mathbf{j}$

$OC = 15\cos 43° = 11.0$
$CR = 15\sin 43° = 10.2$
$\Rightarrow \overrightarrow{OR} = -11.0\mathbf{i} - 10.2\mathbf{j}$

M1

**EXAMPLE 2**

Find the magnitude and direction of the following vectors.
**a** $p = 2i + 5j$ **b** $q = 3i - 2j$ **c** $r = -i - 2j$

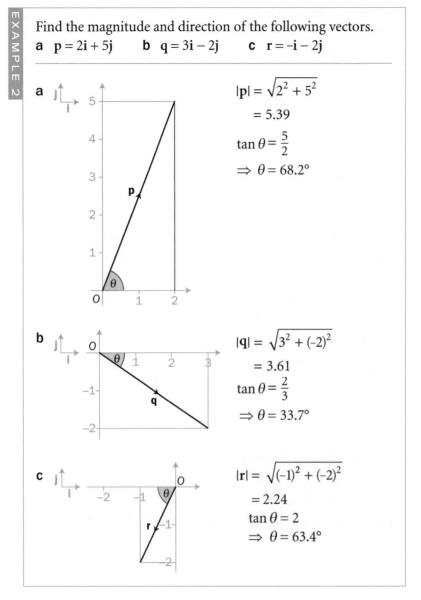

**a**

$$|p| = \sqrt{2^2 + 5^2}$$
$$= 5.39$$
$$\tan\theta = \frac{5}{2}$$
$$\Rightarrow \theta = 68.2°$$

**b**

$$|q| = \sqrt{3^2 + (-2)^2}$$
$$= 3.61$$
$$\tan\theta = \frac{2}{3}$$
$$\Rightarrow \theta = 33.7°$$

**c**

$$|r| = \sqrt{(-1)^2 + (-2)^2}$$
$$= 2.24$$
$$\tan\theta = 2$$
$$\Rightarrow \theta = 63.4°$$

You should sketch diagrams to help you to visualise the vectors. You can then use your diagrams to check if your answers make sense.

The directions here have been described by marking an angle on the diagram and then finding that angle.
More formally, you would give the direction as a rotation through an angle $\theta$ from the positive x-direction, with $-180° < \theta \leqslant 180°$. The anticlockwise sense is positive. These answers would then be a 68.2°, b –33.7°, c –116.6°.

You can find the resultant of vectors easily if they are expressed in component form.

In this diagram
$$p = 3i + j, q = 2i + 3j$$

You can see that
$$p + q = 5i + 4j$$

Or, using column

vectors, $p + q = \begin{pmatrix} 3 \\ 1 \end{pmatrix} + \begin{pmatrix} 2 \\ 3 \end{pmatrix} = \begin{pmatrix} 5 \\ 4 \end{pmatrix}$

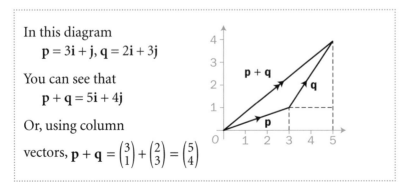

**M1**

When adding vectors, you add the corresponding components. Similarly, if you subtract vectors or multiply a vector by a scalar, you deal with the **i** and **j** components separately.

In general, you have

$$(a\mathbf{i} + b\mathbf{j}) + (c\mathbf{i} + d\mathbf{j}) = (a + c)\mathbf{i} + (b + d)\mathbf{j}$$
$$(a\mathbf{i} + b\mathbf{j}) - (c\mathbf{i} + d\mathbf{j}) = (a - c)\mathbf{i} + (b - d)\mathbf{j}$$
$$k(a\mathbf{i} + b\mathbf{j}) = ka\mathbf{i} + kb\mathbf{j} \quad \text{where } k \text{ is a scalar.}$$

Or, in column vector form,

$$\begin{pmatrix} a \\ b \end{pmatrix} + \begin{pmatrix} c \\ d \end{pmatrix} = \begin{pmatrix} a + c \\ b + d \end{pmatrix}$$

$$\begin{pmatrix} a \\ b \end{pmatrix} - \begin{pmatrix} c \\ d \end{pmatrix} = \begin{pmatrix} a - c \\ b - d \end{pmatrix}$$

$$k\begin{pmatrix} a \\ b \end{pmatrix} = \begin{pmatrix} ka \\ kb \end{pmatrix}$$

**EXAMPLE 3**

Given **p** = 12**i** + 5**j** and **q** = 3**i** − 4**j**, find

**a i** p − q      **ii** 2p + 3q

**b** a vector parallel to **p** with magnitude 39

**c** the unit vector $\hat{\mathbf{q}}$.

- - - - - - - - - -

**a i** $\;$ p − q = 12**i** + 5**j** − (3**i** − 4**j**)
$\qquad = (12 − 3)\mathbf{i} + (5 − (−4))\mathbf{j}$
$\qquad = 9\mathbf{i} + 9\mathbf{j}$

**ii** $\;$ 2p + 3q = 2(12**i** + 5**j**) + 3(3**i** − 4**j**)
$\qquad = (24\mathbf{i} + 10\mathbf{j}) + (9\mathbf{i} − 12\mathbf{j})$
$\qquad = 33\mathbf{i} − 2\mathbf{j}$

**b** $\;|\mathbf{p}| = \sqrt{12^2 + 5^2} = 13$

A vector $k\mathbf{p}$ is parallel to **p** and has $k$ times the magnitude of **p**. The required vector is
$\quad 3\mathbf{p} = 36\mathbf{i} + 15\mathbf{j}$

$39 = 3 \times 13$

**c** $\;|\mathbf{q}| = \sqrt{3^2 + (−4)^2} = 5$

$\hat{\mathbf{q}}$ has a magnitude of 1, so $\hat{\mathbf{q}} = \frac{1}{5}\mathbf{q} = 0.6\mathbf{i} − 0.8\mathbf{j}$

$1 = \frac{1}{5} \times 5$

Example 3 part **c** shows that to find a unit vector you divide the given vector by its magnitude.

$$\hat{\mathbf{a}} = \frac{\mathbf{a}}{|\mathbf{a}|}$$

M1

To find the resultant of vectors defined by their magnitude and direction, it is easier to first convert them to component form.

EXAMPLE 4

Find the magnitude and direction of the resultant of the vectors shown in the diagram.

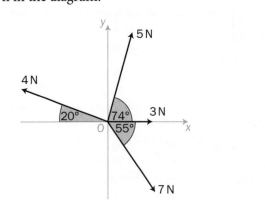

To find the resultant, **R**, express each vector in component form and then add the vectors:

$$\mathbf{R} = \begin{pmatrix} 3 \\ 0 \end{pmatrix} + \begin{pmatrix} 5\cos 74° \\ 5\sin 74° \end{pmatrix} + \begin{pmatrix} -4\cos 20° \\ 4\sin 20° \end{pmatrix} + \begin{pmatrix} 7\cos 55° \\ -7\sin 55° \end{pmatrix}$$

$$= \begin{pmatrix} 4.63 \\ 0.44 \end{pmatrix}$$

You should sketch a diagram of the vector **R**.

Next find the magnitude and direction of **R**:

$$|\mathbf{R}| = \sqrt{4.63^2 + 0.44^2}$$
$$= 4.65$$

$$\tan \theta = \frac{0.44}{4.63}$$

$$\Rightarrow \theta = 5.43°$$

The resultant vector has magnitude 4.65 units and direction 5.43°.

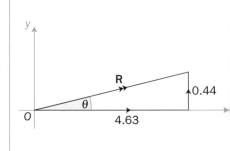

You can split an equation involving two-dimensional vectors into two separate equations, one using only the $x$-components and the other using only the $y$-components.

Two vectors are equal if and only if their corresponding components are equal.
So if $\mathbf{p} = a\mathbf{i} + b\mathbf{j}$ and $\mathbf{q} = c\mathbf{i} + d\mathbf{j}$, then

$$\mathbf{p} = \mathbf{q} \qquad \Leftrightarrow \qquad a = c \quad \text{and} \quad b = d$$

MI

Given that $\mathbf{p} = m\mathbf{i} + n\mathbf{j}$ and $\mathbf{q} = (2n + 5)\mathbf{i} + (1 - m)\mathbf{j}$, find the values of $m$ and $n$ for which $\mathbf{p} = \mathbf{q}$.

You have $\quad m\mathbf{i} + n\mathbf{j} = (2n + 5)\mathbf{i} + (1 - m)\mathbf{j}$

Equate components:

$$m = 2n + 5 \qquad [1]$$
$$n = 1 - m \qquad [2]$$

Substitute from [2] into [1]:

$$m = 2(1 - m) + 5$$

$$\Rightarrow m = 2\frac{1}{3}$$

Substitute back into [2]:

$$n = -1\frac{1}{3}$$

## Exercise 3.2

1 Given vectors $\mathbf{p} = 2\mathbf{i} - \mathbf{j}$, $\mathbf{q} = -2\mathbf{i} + 3\mathbf{j}$ and $\mathbf{r} = 4\mathbf{i} + \mathbf{j}$, calculate

   a  $\mathbf{p} + \mathbf{q}$              b  $\mathbf{p} - \mathbf{r}$

   c  $2\mathbf{q} - \mathbf{p}$           d  $2\mathbf{p} + 3\mathbf{r}$

   e  $|\mathbf{p}|$                 f  $|\mathbf{q} + \mathbf{r}|$

2 Given vectors $\mathbf{p} = 3\mathbf{i} + u\mathbf{j}$, $\mathbf{q} = v\mathbf{i} - 4\mathbf{j}$ and $\mathbf{r} = 4\mathbf{i} - 6\mathbf{j}$, find

   a  the values of $u$ and $v$ if $\mathbf{p} - \mathbf{q} = \mathbf{r}$

   b  the value of $u$ if $\mathbf{p}$ and $\mathbf{r}$ are parallel.

3 Given $\mathbf{p} = -3\mathbf{i} + 4\mathbf{j}$, find

   a  a vector parallel to $\mathbf{p}$ and with magnitude 20

   b  the unit vector $\hat{\mathbf{p}}$ in the direction of $\mathbf{p}$.

4 Find the values of $x$ and $y$ which satisfy these equations.

   a  $(2x + 1)\mathbf{i} + (x - 2)\mathbf{j} = (y - 1)\mathbf{i} + (2 - y)\mathbf{j}$

   b  $(x^2 - y^2)\mathbf{i} + 2xy\mathbf{j} = 3\mathbf{i} + 4\mathbf{j}$

5 Given that $\mathbf{p} = (x - 3)\mathbf{i} + (y + 2)\mathbf{j}$ and $\mathbf{q} = (y + 3)\mathbf{i} + (3x - 2)\mathbf{j}$, that $\mathbf{p}$ and $\mathbf{q}$ are parallel and that $|\mathbf{q}| = 2|\mathbf{p}|$, find the values of $x$ and $y$.

6 Express each of these vectors in the form $x\mathbf{i} + y\mathbf{j}$.

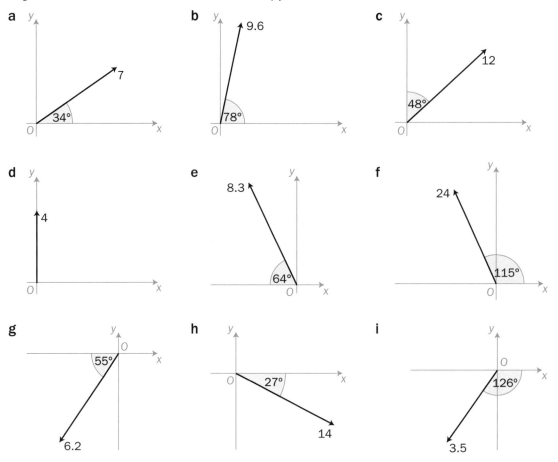

7 Find the magnitude, $r$, and the direction, $\theta$, of these vectors, where $\theta$ is the rotation from the positive $x$-direction and $-180° < \theta \leqslant 180°$.

The positive direction of rotation is anticlockwise.

a $5\mathbf{i} + 2\mathbf{j}$    b $-5\mathbf{j}$

c $-2\mathbf{i} + 3\mathbf{j}$    d $3\mathbf{i} - 5\mathbf{j}$

e $-6\mathbf{i} - 5\mathbf{j}$    f $-2\mathbf{i}$

8 A ship sails from $O$ to $A$, a distance of 20 km on a bearing of 072°, and then from $A$ to $B$, a distance of 28 km on a bearing of 024°. Take east and north to be the $x$- and $y$-directions respectively.

a Express the displacement vectors $\overrightarrow{OA}$ and $\overrightarrow{AB}$ in component form.

b Find the resultant displacement $\overrightarrow{OB}$ in component form.

c Find the magnitude and direction of $\overrightarrow{OB}$.

9 Two ships, $A$ and $B$, leave harbour at $O$. Ship $A$ travels 35 km on a bearing of 280°. Ship $B$ travels 50 km on a bearing of 030°.

Take east and north to be the $x$- and $y$-directions respectively.

a Express the displacement vectors $\overrightarrow{OA}$ and $\overrightarrow{OB}$ in component form.

b Express the vector $\overrightarrow{AB}$ in terms of $\overrightarrow{OA}$ and $\overrightarrow{OB}$, and hence find $\overrightarrow{AB}$ in component form.

c How far and on what bearing is ship $B$ from ship $A$?

10 An aircraft is flying in a wind of speed 50 km h$^{-1}$ from the south west. The aircraft is steered so that it travels at a speed of 300 km h$^{-1}$ on a bearing of 110°.

a Taking east and north as the $\mathbf{i}$- and $\mathbf{j}$-directions respectively, express the velocity of the wind and the velocity of the aircraft in component form. Hence find the speed of the aircraft in still air and the direction in which it is being steered.

b The wind suddenly reverses, so that it blows at 50 km h$^{-1}$ from the north-east.

In what direction and at what speed does the aircraft now travel if it continues to be steered in the same direction?

11 An aircraft, whose speed in still air is $v$ m s$^{-1}$, travels from a point $A$ to a point $B$ and back again. There is a wind of speed $w$ m s$^{-1}$ blowing in a direction making an angle $\theta$ to $AB$.

a Show that the angle between the direction the aircraft steers and the direction in which it travels is the same for both stages of the journey.

b Show that the journey is only possible if $v > w \sin \theta$

M1

The vector $\overrightarrow{OA}$ describes the position of a point $A$ relative to an origin $O$. $\overrightarrow{OA}$ is the position vector of $A$.

Instead of saying that point $A$ has coordinates $(a, b)$, you say that $A$ has position vector $a\mathbf{i} + b\mathbf{j}$.

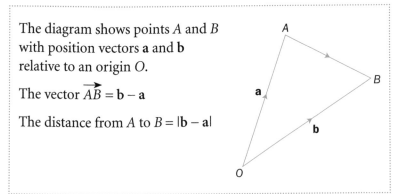

The diagram shows points $A$ and $B$ with position vectors $\mathbf{a}$ and $\mathbf{b}$ relative to an origin $O$.

The vector $\overrightarrow{AB} = \mathbf{b} - \mathbf{a}$

The distance from $A$ to $B = |\mathbf{b} - \mathbf{a}|$

**EXAMPLE 1**

Points $A$ and $B$ have position vectors $\mathbf{a} = 2\mathbf{i} + \mathbf{j}$ and $\mathbf{b} = 5\mathbf{i} - 6\mathbf{j}$ respectively. Find the distance $AB$.

$$\overrightarrow{AB} = \mathbf{b} - \mathbf{a}$$
$$= (5\mathbf{i} - 6\mathbf{j}) - (2\mathbf{i} + \mathbf{j}) = 3\mathbf{i} - 7\mathbf{j}$$

Find the magnitude:

$$AB = |\mathbf{b} - \mathbf{a}|$$
$$= \sqrt{3^2 + (-7)^2} = \sqrt{58} = 7.62$$

For a point moving in two dimensions with constant velocity, its velocity vector tells you its displacement in each unit of time.

**EXAMPLE 2**

A particle leaves point $A$, with position vector $(3\mathbf{i} + 7\mathbf{j})$ m, and travels with constant velocity $(2\mathbf{i} - \mathbf{j})$ m s$^{-1}$. Find its position, $B$, after 3 s.

$$\overrightarrow{OA} = 3\mathbf{i} + 7\mathbf{j}$$

The change of position in 3s is
$$\overrightarrow{AB} = 3 \times (2\mathbf{i} - \mathbf{j}) = 6\mathbf{i} - 3\mathbf{j}$$

The position vector of $B$ is $\overrightarrow{OB}$.
$$\overrightarrow{OB} = \overrightarrow{OA} + \overrightarrow{AB}$$
$$\overrightarrow{OB} = (3\mathbf{i} + 7\mathbf{j}) + (6\mathbf{i} - 3\mathbf{j}) = 9\mathbf{i} + 4\mathbf{j}$$

So the position vector of $B$ is $(9\mathbf{i} + 4\mathbf{j})$ m.

Always sketch a diagram.

M1

EXAMPLE 3

At a certain time, particle $A$ is at the point with position vector $(\mathbf{i} + 4\mathbf{j})$ m and is moving with constant velocity $(3\mathbf{i} + 3\mathbf{j})$ m s$^{-1}$. At the same time, particle $B$ is at the point $(5\mathbf{i} + 2\mathbf{j})$ m and is moving with constant velocity $(2\mathbf{i} + 3.5\mathbf{j})$ m s$^{-1}$.

**a** Find the vector $\overrightarrow{AB}$ at time $t$ s.

**b** Hence show that the particles will collide, and find the position vector of the point of collision.

**a** The position of $A$ changes by $(3\mathbf{i} + 3\mathbf{j})$ m every second.
$A$ starts at $(\mathbf{i} + 4\mathbf{j})$ m.
After $t$ seconds,

$$\overrightarrow{OA} = (\mathbf{i} + 4\mathbf{j}) + (3\mathbf{i} + 3\mathbf{j})t$$
$$= (1 + 3t)\mathbf{i} + (4 + 3t)\mathbf{j}$$

Similarly, after $t$ seconds,

$$\overrightarrow{OB} = (5\mathbf{i} + 2\mathbf{j}) + (2\mathbf{i} + 3.5\mathbf{j})t$$
$$= (5 + 2t)\mathbf{i} + (2 + 3.5t)\mathbf{j}$$

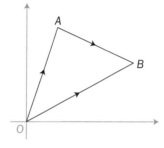

You can see from the diagram that $\overrightarrow{AB} = \overrightarrow{OB} - \overrightarrow{OA}$.
So

$$\overrightarrow{AB} = ((5 + 2t)\mathbf{i} + (2 + 3.5t)\mathbf{j}) - ((1 + 3t)\mathbf{i} + (4 + 3t)\mathbf{j})$$
$$= (4 - t)\mathbf{i} + (0.5t - 2)\mathbf{j}$$

**b** The particles will collide if there is a value of $t$ for which $\overrightarrow{AB} = \mathbf{0}$.
The $x$- and $y$-components of $\overrightarrow{AB}$ must be zero for the same value of $t$.

Equate the $x$-component to zero: $\qquad 4 - t = 0$
$\qquad\qquad\qquad \Rightarrow \qquad\qquad t = 4$

Equate the $y$-component to zero: $\qquad 0.5t - 2 = 0$
$\qquad\qquad\qquad \Rightarrow \qquad\qquad t = 4$

So the particles collide when $t = 4$.

At $t = 4$, particle $A$ has position vector
$$\overrightarrow{OA} = (1 + 3 \times 4)\mathbf{i} + (4 + 3 \times 4)\mathbf{j}$$
$$= 13\mathbf{i} + 16\mathbf{j}$$

Check that $B$ has the same position vector at $t = 4$:
$$\overrightarrow{OB} = (5 + 2 \times 4)\mathbf{i} + (2 + 3.5 \times 4)\mathbf{j}$$
$$= 13\mathbf{i} + 16\mathbf{j}$$
So the particles collide at the point with position vector $13\mathbf{i} + 16\mathbf{j}$.

M1

For a point moving in two dimensions with constant acceleration, its acceleration vector tells you the change of velocity in each unit of time.

EXAMPLE 4

A particle moving with velocity $(\mathbf{i} + 2\mathbf{j})$ m s$^{-1}$ undergoes a constant acceleration of $(2\mathbf{i} - \mathbf{j})$ m s$^{-2}$ for a period of 4 s. Find its velocity and speed at the end of this period.

The change of velocity $= 4 \times (2\mathbf{i} - \mathbf{j}) = (8\mathbf{i} - 4\mathbf{j})$ m s$^{-1}$

The new velocity $= (\mathbf{i} + 2\mathbf{j}) + (8\mathbf{i} - 4\mathbf{j}) = (9\mathbf{i} - 2\mathbf{j})$ m s$^{-1}$

Speed is the magnitude of velocity

$\Rightarrow$ Speed $= |9\mathbf{i} - 2\mathbf{j}| = \sqrt{9^2 + (-2)^2} = \sqrt{85} = 9.22$ m s$^{-1}$

In effect, this problem has used $\mathbf{v} = \mathbf{u} + \mathbf{a}t$ as a vector equation. You will recognise this equation from Chapter 2. This is only valid for constant acceleration.

EXAMPLE 5

Take east and north to be the $x$- and $y$-directions respectively. A bird, flying with velocity $(3\mathbf{i} - 4\mathbf{j})$ m s$^{-1}$, accelerates with constant acceleration $(\mathbf{i} + 2\mathbf{j})$ m s$^{-2}$.

**a** At what subsequent time is the bird flying due east?

**b** Find the bird's speed and direction of flight after 5 s.

**a** After $t$ s, the bird's velocity has changed by $(\mathbf{i} + 2\mathbf{j})t$ m s$^{-1}$.
Hence velocity at time $t$ s $= (3\mathbf{i} - 4\mathbf{j}) + (\mathbf{i} + 2\mathbf{j})t$
$= (3 + t)\mathbf{i} + (2t - 4)\mathbf{j}$ m s$^{-1}$

The bird is flying due east when the $y$-component of its velocity is zero.

Equate the $y$-component to zero: $2t - 4 = 0$
$\Rightarrow \qquad t = 2$

So the bird is flying due east after 2 seconds.

**b** After 5 s, velocity $= (3 + 5)\mathbf{i} + (2 \times 5 - 4)\mathbf{j}$
$= (8\mathbf{i} + 6\mathbf{j})$ m s$^{-1}$

Its speed $v = |8\mathbf{i} + 6\mathbf{j}|$
$= \sqrt{8^2 + 6^2}$
$= 10$ m s$^{-1}$

Its direction is given by $\theta$, where

$\tan \theta = \dfrac{6}{8}$

$\Rightarrow \qquad \theta = 36.9°$

The bird is flying on a bearing of $053.1°$

## Exercise 3.3

1  Points $A$ and $B$ have position vectors $(-2\mathbf{i} + 3\mathbf{j})$ m and $(4\mathbf{i} + 7\mathbf{j})$ m respectively. Find

   a   the length $AB$

   b   the angle made by $AB$ with the $x$-direction.

2  A particle, $A$, is at rest at the point with position vector $(3\mathbf{i} + \mathbf{j})$ m. A second particle, $B$, starts from the origin and moves with constant velocity $(2\mathbf{i} + 3\mathbf{j})$ m s$^{-1}$. Find

   a   the position vector of $B$ after 3 s

   b   the distance and direction of $B$ from $A$ at this time.

3  A particle is moving with constant velocity $(5\mathbf{i} + 3\mathbf{j})$ m s$^{-1}$. Find

   a   the speed of the particle

   b   the angle between the path of the particle and the $x$-direction.

4  A particle, moving with velocity $(-\mathbf{i} + 2\mathbf{j})$ m s$^{-1}$, undergoes a constant acceleration of $(2\mathbf{i} + \mathbf{j})$ m s$^{-2}$ for 5 s. Find

   a   the speed of the particle at the end of this period

   b   the direction in which the particle is moving at this time.

5  A particle, moving with velocity $(6\mathbf{i} - \mathbf{j})$ m s$^{-1}$, undergoes a constant acceleration of $(3\mathbf{i} - 2\mathbf{j})$ m s$^{-2}$.

   a   Find its velocity at time $t$ s.

   b   Show that at no time does it travel parallel to either axis.

   c   Find the time at which it has a speed of 13 m s$^{-1}$.

6  A particle starts from the point with position vector $4\mathbf{j}$ m and moves with constant velocity $(2\mathbf{i} - \mathbf{j})$ m s$^{-1}$. At the same time a second particle starts from the point with position vector $(6\mathbf{i} + 8\mathbf{j})$ m and moves with constant velocity $(-\mathbf{i} - 3\mathbf{j})$ m s$^{-1}$. Show that the particles collide, and find the time at which they do so.

M1

7 A particle $A$ starts from the point with position vector $5\mathbf{j}$ m and moves with constant velocity $(2\mathbf{i} + \mathbf{j})$ m s$^{-1}$. Five seconds later a second particle, $B$, leaves the origin moving with constant velocity $(p\mathbf{i} + q\mathbf{j})$ m s$^{-1}$. Given that particle $B$ collides with particle $A$ after a further 2 s, find the values of $p$ and $q$.

8 Starting simultaneously from the same point, $O$, Alvin and Bernard set out across a field in the fog. Their velocities are constant and are $(\mathbf{i} + 2\mathbf{j})$ m s$^{-1}$ and $(3\mathbf{i} + \mathbf{j})$ m s$^{-1}$ respectively. To avoid losing each other they hold opposite ends of a 90 m string.

   a At what speed is Alvin travelling?

   b Find the directions of travel of the two people, and hence the angle between their paths.

   c Find the position vectors $\overrightarrow{OA}$ and $\overrightarrow{OB}$ after a time of $t$ seconds, and hence the vector $\overrightarrow{AB}$ at this time.

   d Find the value of $t$ for which the string becomes taut.

9 Take east and north to be the $x$- and $y$-directions respectively, and the harbour to be the origin.
   A ship, $S$, leaves the harbour travelling with constant velocity $(5\mathbf{i} + 10\mathbf{j})$ km h$^{-1}$. There is a lighthouse, $L$, at the point with position vector $(30\mathbf{i} + 20\mathbf{j})$ km.

   a Write down the position vector of the ship at time $t$ hours after setting out.

   b Write down the vector $\overrightarrow{LS}$ at time $t$.

   c Find the time at which the ship is due west of the lighthouse.

   d Find the distance $d$ between the ship and the lighthouse at the time found in part c.

   e Find the time at which the ship is again a distance $d$ from the lighthouse.

10 Take east and north to be the $x$- and $y$-directions respectively.
   Ship $A$ starts from the point with position vector $(2\mathbf{i} + 3\mathbf{j})$ km and sails at a constant velocity of $(2\mathbf{i} + \mathbf{j})$ km h$^{-1}$.
   Simultaneously ship $B$ leaves the point with position vector $(4\mathbf{i} + \mathbf{j})$ km and sails at a constant velocity of $(3\mathbf{i} + 4\mathbf{j})$ km h$^{-1}$.
   Find the shortest distance between the two ships in the subsequent motion.

11 Two spiders, *A* and *B*, are in a field. Their position vectors relative to an origin *O* are $(\mathbf{i} + 3\mathbf{j})$ m and $(4\mathbf{i} + 2\mathbf{j})$ m. They start to run at the same time, *A* with constant velocity $(3\mathbf{i} + \mathbf{j})$ m s$^{-1}$ and *B* with constant velocity $(\mathbf{i} + 2\mathbf{j})$ m s$^{-1}$. At time *t* s their position vectors are **a** and **b** respectively.

   **a**  Write down the vectors **a** and **b** in terms of *t*.

   **b**  Write down the vector $\overrightarrow{AB}$ in terms of *t*.

   **c**  Show that the spiders do not meet.

   **d**  Find the distance between the spiders when *t* = 3 s.

At time *t* = 3 s *B* decides to run in the **i**-direction at constant velocity *v* m s$^{-1}$ in order to meet up with *A*. Assuming that *A* does not change its velocity

   **e**  find the value of *v*.

12 Take the harbour, *O*, to be the origin, and east and north to be the *x*- and *y*-directions respectively.

A ship, *P*, starts at midday from a point 20 km north of *O* and sails with constant velocity $(3\mathbf{i} - 2\mathbf{j})$ km h$^{-1}$. A second ship, *Q*, leaves *O* at midday and sails with constant velocity $(6\mathbf{i} + 2\mathbf{j})$ km h$^{-1}$. The position vectors of the ships at time *t* hours after midday are **p** and **q** respectively.

   **a**  Express **p** and **q** in terms of *t*.

   **b**  Express the vector $\overrightarrow{PQ}$ in terms of *t*.

   **c**  If the distance between the ships at time *t* hours is *d* km, show that

$$d^2 = 25t^2 - 160t + 400$$

   **d**  By completing the square, or otherwise, show that the minimum distance between the ships is 12 km, and find the value of *t* for which this occurs.

M1

1 Points $A$ and $B$ are directly opposite each other across a river which is 100 m wide and flowing at 2 m s$^{-1}$. A boat, which can travel at 4 m s$^{-1}$ in still water, leaves $A$ to cross the river.

    **a**   If the boat is steered directly across the river, how far downstream of $B$ will it reach the other bank?

    **b**   In what direction should it be steered so that it travels directly to $B$? How long will the trip then take?

2 Given two vectors, $(3\mathbf{i} + 2\mathbf{j})$ and $(2\mathbf{i} - \mathbf{j})$, find

    **a**   the magnitude of the resultant vector

    **b**   the angle which the resultant vector makes with the direction of $\mathbf{i}$.

3 Two parties of walkers, $A$ and $B$, set out from the same camp at the same time. Party $A$ walks at 4 km h$^{-1}$ on a bearing of 030°. Party $B$ walks due east at 5 km h$^{-1}$. Take the camp to be the origin, and take east and north to be the $\mathbf{i}$- and $\mathbf{j}$-directions.

    **a**   Find the position vectors of both parties after 3 hours, in the form $(p\mathbf{i} + q\mathbf{j})$.

    **b**   Find the distance and bearing of party $B$ from party $A$ at this time.

4 Two boats, $A$ and $B$, are moving with constant velocities. Boat $A$ moves with velocity $9\mathbf{j}$ km h$^{-1}$. Boat $B$ moves with velocity $(3\mathbf{i} + 5\mathbf{j})$ km h$^{-1}$.

*The horizontal unit vectors $\mathbf{i}$ and $\mathbf{j}$ are directed due east and due north respectively.*

    **a**   Find the bearing on which $B$ is moving.

At noon, $A$ is at point $O$, and $B$ is 10 km due west of $O$. At time $t$ hours after noon, the position vectors of $A$ and $B$ relative to $O$ are $\mathbf{a}$ km and $\mathbf{b}$ km respectively.

    **b**   Find expressions for $\mathbf{a}$ and $\mathbf{b}$ in terms of $t$, giving your answer in the form $p\mathbf{i} + q\mathbf{j}$.

    **c**   Find the time when $B$ is due south of $A$.

At time $t$ hours after noon, the distance between $A$ and $B$ is $d$ km. By finding an expression for $\overrightarrow{AB}$

    **d**   show that $d^2 = 25t^2 - 60t + 100$.

At noon, the boats are 10 km apart.

    **e**   Find the time after noon at which the boats are again 10 km apart.   [(c) Edexcel Limited 2004]

**5** A particle, $P$, is moving with constant acceleration. At $t = 0$, $P$ has velocity $(3\mathbf{i} - 5\mathbf{j})$ m s$^{-1}$. At $t = 4$ s, the velocity of $P$ is $(-5\mathbf{i} + 11\mathbf{j})$ m s$^{-1}$. Find

    **a** the acceleration of $P$, in terms of $\mathbf{i}$ and $\mathbf{j}$.

At $t = 6$ s, $P$ is at point $A$ with position vector $(6\mathbf{i} - 29\mathbf{j})$ m relative to a fixed origin $O$. At this instant the acceleration of $P$ ceases and $P$ then moves with constant velocity. Three seconds later, $P$ reaches point $B$.

    **b** Calculate the distance of $B$ from $O$.

**6** A boat which has a speed of 10 m s$^{-1}$ in still water is being steered due north. There is a strong current flowing, and the boat actually travels at a speed of 18 m s$^{-1}$ on a bearing of 060°.

    **a** Taking east and north as the $\mathbf{i}$- and $\mathbf{j}$-directions, calculate the velocity of the current in component form.

    **b** Hence calculate the speed and direction of the current.

    **c** Explain, by means of a vector diagram, why the boat cannot return to its starting point.

**7** At time $t = 0$, a football player kicks a ball from the point $A$ with position vector $(2\mathbf{i} + \mathbf{j})$ m on a horizontal football field. The motion of the ball is modelled as that of a particle moving horizontally with constant velocity $(5\mathbf{i} + 8\mathbf{j})$ m s$^{-1}$. Find

*In this question, the unit vectors $\mathbf{i}$ and $\mathbf{j}$ are horizontal vectors due east and north respectively.*

    **a** the speed of the ball

    **b** the position vector of the ball after $t$ seconds.

The point $B$ on the field has position vector $(10\mathbf{i} + 7\mathbf{j})$ m.

    **c** Find the time when the ball is due north of $B$.

At time $t = 0$, another player starts running due north from $B$ and moves with constant speed $v$ m s$^{-1}$. Given that he intercepts the ball

    **d** find the value of $v$.

    **e** State one physical factor, other than air resistance, which would be needed in a refinement of the model of the ball's motion to make the model more realistic.

[(c) Edexcel Limited 2005]

M1

# 3

## Exit ⟹

Summary

- A vector has magnitude and direction.
  - Vectors are equal if they have the same magnitude and the same direction.
  - $\overrightarrow{AB} + \overrightarrow{BC} = \overrightarrow{AC}$ is the vector sum or the resultant vector of $\overrightarrow{AB}$ and $\overrightarrow{BC}$.
  - $k\mathbf{a}$ is a vector parallel to $\mathbf{a}$ and with magnitude $k$ times greater than $|\mathbf{a}|$.
  - Subtracting a vector is the same as adding its negative:
    $$\mathbf{p} - \mathbf{q} = \mathbf{p} + (-\mathbf{q})$$
- $\mathbf{i}$ and $\mathbf{j}$ are the unit vectors in the $x$- and $y$ directions.
  - You write a vector in component form as
    $$(x\mathbf{i} + y\mathbf{j}) \quad \text{or} \quad \begin{pmatrix} x \\ y \end{pmatrix}$$
  - Given magnitude $r$ and direction $\theta$, you can resolve the vector into components as $\overrightarrow{OP} = r\cos\theta\,\mathbf{i} + r\sin\theta\,\mathbf{j}$
  - Given components $x$ and $y$, you can find magnitude and direction using
    $$r = \sqrt{x^2 + y^2} \text{ and } \tan\theta = \frac{y}{x}$$
  - $(a\mathbf{i} + b\mathbf{j}) + (c\mathbf{i} + d\mathbf{j}) = (a + c)\mathbf{i} + (b + d)\mathbf{j}$
    $(a\mathbf{i} + b\mathbf{j}) - (c\mathbf{i} + d\mathbf{j}) = (a - c)\mathbf{i} + (b - d)\mathbf{j}$
    $k(a\mathbf{i} + b\mathbf{j}) = ka\mathbf{i} + kb\mathbf{j}$ where $k$ is a scalar
  - If $\mathbf{p} = a\mathbf{i} + b\mathbf{j}$ and $\mathbf{q} = c\mathbf{i} + d\mathbf{j}$, then $\mathbf{p} = \mathbf{q} \Leftrightarrow a = c$ and $b = d$
- The position vector of a point is its displacement from the origin.
  - If points $A$ and $B$ have position vectors $\mathbf{a}$ and $\mathbf{b}$, then $\overrightarrow{AB} = \mathbf{b} - \mathbf{a}$
  - If a point has a constant velocity vector $\mathbf{v}$, its position vector changes by $\mathbf{v}t$ in time $t$.
  - If a point has a constant acceleration vector $\mathbf{a}$, its velocity vector changes by $\mathbf{a}t$ in time $t$.

Refer to

3.1

3.2

3.3

Always sketch a diagram to visualise the vectors.

MI

---

### Links

Vectors are used in navigation. Some important factors to consider when deciding on a flight route are the type of aircraft, the weather conditions and, perhaps most importantly, the flight distance. It is often important that the shortest route is chosen. This is however not as simple as finding the straight line between two points. The Earth's surface is curved and so the shortest route between two points on the globe is in fact an arc. These routes can be found using vectors.

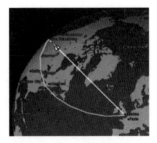

# 4

# Forces

This chapter will show you how to
- identify the forces involved in practical situations
- model forces as vectors
- use a variety of methods to solve problems involving forces in equilibrium
- use the standard model of friction.

## Before you start

### You should know how to:

1 Solve problems using Pythagoras' theorem and the trigonometry of right-angled triangles.

2 Solve linear simultaneous equations.

3 Manipulate vectors in two dimensions.

### Check in

1 Find the angle $x$ and the side length $y$ in this diagram.

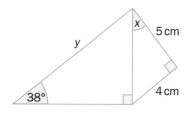

2 Solve the simultaneous equations

$$3x - 4y = 2$$
$$5x + 6y = 35$$

3

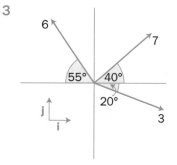

a Find in component form the resultant of the vectors shown.
b Find the magnitude and direction of the resultant.

When you lift an object, you feel the force of gravity.
If you take part in a tug-of-war, you feel the tension as a force in the rope, and the frictional force stops your feet from slipping.
When you kneel on the floor, you feel a reaction as the force of the floor pressing on your knees.

All of these different forces are vector quantities, because they have both a magnitude and a direction.

In this chapter you will deal mostly with statics, i.e. with forces acting on a stationary object. The resultant of the forces must be zero, or the object would move. The forces are in equilibrium.

The SI unit of force is the newton (N).

You need to be clear about the difference between mass and weight.

Their effect depends on where they are applied – their **line of action**.
e.g. Lifting a ladder at the middle or at the end has different effects. This notion is covered in 7.1.
In the present chapter, objects are assumed to be particles, so there is only one point at which the force can act.

A force of 1 N will give a mass of 1 kg an acceleration of $1 \text{ m s}^{-2}$.
Refer to 5.1 for more detail.

The mass of an object
- is the amount of matter forming the object
- is measured in kilograms
- is the same wherever in the universe the object is placed.

Compare with

The weight of an object
- is the force acting on the object in a gravitational field
- is measured in newtons
- is different for the same object in different positions.
  e.g. The weight of an object is different when it is on the Earth from when it is on the Moon.

Weight and mass are connected by the relationship

weight = mass × acceleration due to gravity

You will cover this relationship in more detail in 5.2.

The acceleration due to gravity near the Earth's surface is

$$g \approx 9.8 \text{ m s}^{-2}$$

For an object of mass $m$ kg near the surface of the Earth

weight, $w = mg = 9.8m$ N

In addition to weight you will meet other main types of force – tension, thrust, normal reaction and friction.

- A force due to tension occurs in a string when the ends are pulled.
- If the string is light, the force due to tension is the same throughout the string.

Light means you assume that the mass of the string is negligible compared to the rest of the system.

The woman is holding a light string, from which hangs an object of mass 3 kg.

The tension, $T$, exerts an upward force on the object, and an equal downward force on the woman's hand.

If the object is stationary, the tension must balance the weight, so

$$T = 3g \text{ N}$$

The woman's hand is stationary, so she must be exerting an upward force of $3g$ N to balance the downward pull of the tension in the string.

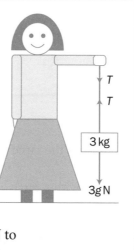

You can assume that the tension in a light string passing over a pulley is constant throughout its length provided that the pulley is smooth.

Smooth means you assume that the friction forces are negligible compared to the other forces.

M1

EXAMPLE 1

Two boxes, $A$ and $B$, of mass 40 kg and 50 kg respectively, are connected by a light rope. A second rope, attached to $B$, passes over a smooth pulley and is fixed to the ground at $C$. Find the forces acting on

**a** the box $A$     **b** the box $B$     **c** the ground at $C$.

**a** The forces on $A$ are
its weight, $40g$ N,
acting downwards
and the tension $T_1$
acting upwards.
The box is stationary,
so     $T_1 = 40g$ N

**b** The forces on $B$ are
its weight, $50g$ N,
the tension $T_1$
acting downwards
and the tension
$T_2$ acting upwards.
The box is stationary,
so     $T_2 = T_1 + 50g = 90g$ N

**c** The pulley is smooth, so the tension is the same throughout the rope joining $B$ to $C$.
So the rope is pulling at the ground with a force of $90g$ N.

Results are often left in terms of $g$.
You could have calculated
$T_1 = 40 \times 9.8 = 392$ N

Draw a diagram to show all the forces acting.
Listing the relevant forces and their direction for each question part will help you to structure your working and ensure you don't leave anything out.

- A **thrust** or **compression** occurs in a rod when the ends are pushed.
- If the rod is **light**, the thrust is the same throughout.

You can also pull a rod, so a rod may be in tension.
You can pull a string but you can not push a string. So a string can have a tension but not a thrust.

Here is a bird table supported by a light, vertical rod.

The rod exerts an upward thrust, $T$, on the table, equal to the downward force $W$, which is the weight of the table and the bird.

The rod exerts a downward thrust force on the ground.

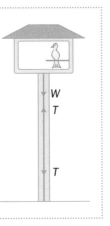

Forces, sometimes called contact forces, occur when two surfaces are touching.

> o A normal reaction occurs when two surfaces are pressed against one another.
> o A normal reaction always acts at right angles to (normal to) the plane of contact.

The diagram shows a ladder leaning against a fence. The ladder is tied to the fence to stop it from slipping.

There is a normal reaction $R_1$ where the ladder is in contact with the ground.

There is a normal reaction $R_2$ where the ladder is in contact with the top of the fence.

The other forces on the ladder are its weight and the tension in the string tying it to the fence and possibly friction too.

> o A frictional force occurs when two surfaces are sliding or tending to slide over one another.
> o The friction always acts to try to prevent the motion.

As the pulling force, $P$, increases, the friction force, $F$, increases to balance it, so that the box does not move.

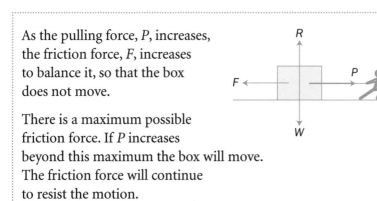

There is a maximum possible friction force. If $P$ increases beyond this maximum the box will move. The friction force will continue to resist the motion.

The maximum frictional force depends on the normal reaction force, $R$, and on the nature of the surfaces.

In some cases the maximum frictional force is small enough to be ignored – the contact is smooth.

You will cover modelling friction in 4.3.

You also get resistive forces when an object moves through the air or through a liquid. Like contact friction, these resistances always act to oppose the motion of the object.

## Drawing diagrams

When solving problems in mechanics you need to draw clear diagrams, showing all the forces and the relevant lengths and angles.

The diagrams should be carefully drawn and of a good size. They will help you to analyse the problem and to explain clearly the steps in your solution.

EXAMPLE 2

A large box of mass $m$ kg is being towed up a rough slope inclined at 30° to the horizontal, using a rope at 20° to the slope. Draw a diagram to show the forces acting on the box.

The forces acting on the box are:
- its weight, $mg$
- the tension, $T$, in the rope
- the normal reaction, $R$
- the friction, $F$, which acts down the slope to oppose the motion.

Make sure you include all of the forces in your diagram.

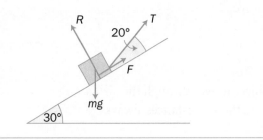

EXAMPLE 3

The box in Example 2 is now allowed to slide down the slope, controlled by the rope.
Draw the forces in this situation.

The only difference is that the box is now moving down the slope, so the frictional force acts up the slope to oppose this motion.

A car of mass $M$ kg is towing a trailer of mass $m$ kg using a light, rigid towbar. There are resistive forces $F_C$ and $F_T$ acting on the car and trailer respectively. Draw diagrams to show the forces acting on the car and on the trailer

**a** when the car exerts a driving force $P$ N

**b** when the car exerts a braking force $B$ N.

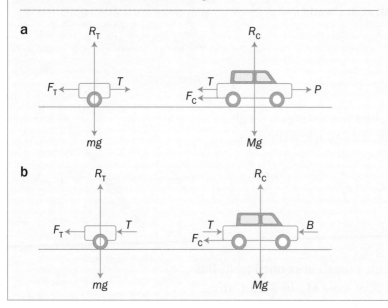

In part **a** the towbar is in tension, and in part **b** it is under thrust.

You only show the tension/thrust in the diagram if you are considering the car and trailer separately. If you take the car and trailer together as one system, the forces in the towbar are **internal forces**, and so you do not show them.

There is actually a separate reaction force at each wheel, but you can usually think of a single, combined normal reaction as shown.

M1

In Example 4 the driving force $P$ is really the friction between each driven wheel and the ground, which acts forward to prevent the wheels from spinning. It is usual to just show it as one generalised driving force.

Similarly, the braking force is really the friction between each wheel and the ground.

## Exercise 4.1

**1** Copy each of these diagrams and mark in the forces indicated.

**a** The forces acting on this brick, which is sliding down a rough inclined plank.

**b** The forces on this uniform shelf, resting symmetrically on two brackets.

**c** The forces on this uniform shelf, resting on one bracket and supported by an inclined wire.

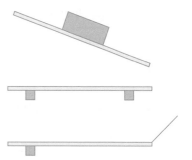

**d** The forces on this ball, which has been thrown vertically, at *A* (on the way up), at *B* (the top of its flight) and at *C* (on the way down).

**e** The forces on this football at *A*, *B* and *C*.

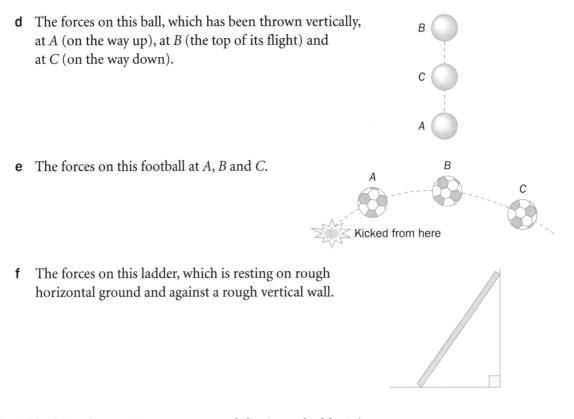

Kicked from here

**f** The forces on this ladder, which is resting on rough horizontal ground and against a rough vertical wall.

**2** A block *A*, of mass $M_A$, rests on a rough horizontal table. It is connected to a second block *B*, of mass $M_B$, by a light string. The string passes over a smooth pulley at the edge of the table and *B* hangs suspended.

**a** Draw a diagram to show the forces acting on each block.

**b** Assuming that the system remains at rest, give the values of the tension in the string and the frictional force acting on block *A*.

**3** A man of mass *m* stands in a lift of mass *M* which is supported by a cable. Draw separate diagrams to show the forces acting

**a** on the lift

**b** on the man.

**4** In this diagram the pulleys are smooth, and the strings are light and inextensible.

**a** Draw a diagram showing the forces acting on the 6 kg mass.

**b** Calculate the frictional force acting on the 6 kg mass if the system is at rest.

5   The diagram shows a light rod *AB* hinged to a vertical
wall at *A*. A light string *BC* is attached to the rod at
*B* and to the wall at *C*. A mass of 2 kg is
suspended from *B*. Draw a diagram showing
the forces acting at the point *B*.

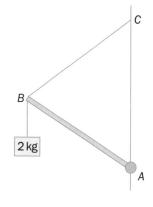

6   The diagram shows a cylinder resting with its axis horizontal.
The cylinder is supported by two rough planes inclined at
50° and 70° to the horizontal. Copy the diagram and
show the forces acting on the cylinder.

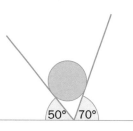

7   A uniform ladder of mass *m* and length 4*a* rests with one end
on rough horizontal ground. The ladder leans against a rough
garden wall so that one quarter of its length protrudes over the
wall. Draw a diagram to show the forces acting on the ladder.

8   A glass rod rests in a smooth hemispherical bowl so that part
of the rod extends beyond the rim of the bowl. Draw a diagram
to show the forces acting on the rod.

9   A small object of mass *m* is suspended at one end of a light
string, the other end of which is tied to a fixed point *A*. The
object moves in a horizontal circle, the centre of which is
vertically below *A*. Draw a diagram to show the forces acting
on the object.

10  A light inextensible string *AB*, of length 1 m, is attached to
the point *B* on the surface of a sphere of radius 1 m. The end
*A* of the string is attached to a point on a vertical wall and the
sphere hangs against the wall.

    **a**  Assuming that the contact between the sphere and the
wall is smooth, draw a diagram of the situation, stating
the angle between the string and the wall and showing the
forces acting on the sphere.

    **b**  Explain how your diagram might be different if the contact
between the sphere and the wall is rough.

M1

# Forces at a point: modelling by vectors

Force is a vector quantity. If several forces act at a point, you can use vector methods to find their combined effect.

EXAMPLE 1

The diagram shows four dog leads *OA*, *OB*, *OC* and *OD*, each tied to the same post *O*. The dogs all pull horizontally on the leads with the forces shown. Find the resultant force on the post.

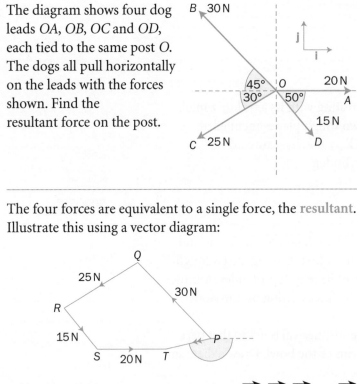

The four forces are equivalent to a single force, the **resultant**. Illustrate this using a vector diagram:

Represent the forces by the displacements $\overrightarrow{PQ}$, $\overrightarrow{QR}$, $\overrightarrow{RS}$ and $\overrightarrow{ST}$.

The displacement vector $\overrightarrow{PT}$ then corresponds to the resultant force.

If you draw the diagram to scale, you will find by measurement that the resultant is approximately 13.5 N acting at approximately 168° to the direction of **i** in the first diagram.

To calculate the resultant, **R**, add the vectors in component form:

$$\mathbf{R} = \begin{pmatrix} 20 \\ 0 \end{pmatrix} + \begin{pmatrix} -30\cos 45° \\ 30\sin 45° \end{pmatrix} + \begin{pmatrix} -25\cos 30° \\ -25\sin 30° \end{pmatrix} + \begin{pmatrix} 15\cos 50° \\ -15\sin 50° \end{pmatrix} = \begin{pmatrix} -13.2 \\ -2.78 \end{pmatrix}$$

The magnitude of **R** is given by

$$|\mathbf{R}| = \sqrt{13.2^2 + 2.78^2} = 13.5 \text{ N}$$

and its direction is $\theta$, where

$$\tan\theta = \frac{2.78}{13.2} \quad \Rightarrow \quad \theta = 11.9°$$

So the resultant force is 13.5 N at an angle of 168.1° below the direction of **i**.

> Methods involving scale drawing are not a requirement of the M1 syllabus and will **not** be accepted as solutions in examinations. You should use them only to check your answer.

In Example 1, the four dogs produced a non-zero resultant force on the post. If you introduced a fifth dog pulling with a force of 13.5 N at an angle of 11.9° to *OA*, the forces would be in equilibrium. The vector diagram would then be a closed polygon.

This additional force is sometimes called the **equilibrant**.

> The vector diagram corresponding to a set of forces in equilibrium is a closed polygon, called the **polygon of forces**.

There are several methods for finding unknown forces when a system is in equilibrium. The first method is by **resolving forces**.

> Concurrent forces are in equilibrium if their resultant is zero. Hence the sum of their components in any direction is zero.

The M1 syllabus does not require you to use a scale drawing to solve a problem. However knowing that the vector diagram is a closed polygon can be useful, especially if there are only three forces. In this case you have a **triangle of forces**.

**Concurrent** forces all act through the same point.

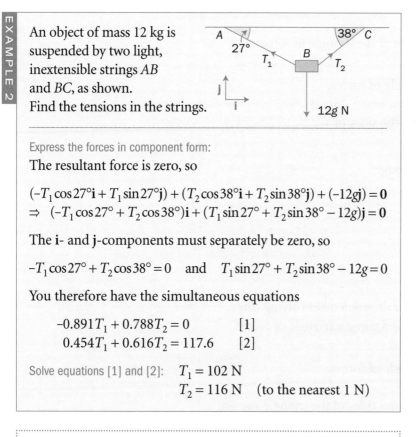

**EXAMPLE 2**

An object of mass 12 kg is suspended by two light, inextensible strings *AB* and *BC*, as shown.
Find the tensions in the strings.

Express the forces in component form:
The resultant force is zero, so

$$(-T_1\cos 27°\mathbf{i} + T_1\sin 27°\mathbf{j}) + (T_2\cos 38°\mathbf{i} + T_2\sin 38°\mathbf{j}) + (-12g\mathbf{j}) = \mathbf{0}$$
$$\Rightarrow (-T_1\cos 27° + T_2\cos 38°)\mathbf{i} + (T_1\sin 27° + T_2\sin 38° - 12g)\mathbf{j} = \mathbf{0}$$

The **i**- and **j**-components must separately be zero, so

$$-T_1\cos 27° + T_2\cos 38° = 0 \quad \text{and} \quad T_1\sin 27° + T_2\sin 38° - 12g = 0$$

You therefore have the simultaneous equations

$$-0.891T_1 + 0.788T_2 = 0 \qquad [1]$$
$$0.454T_1 + 0.616T_2 = 117.6 \qquad [2]$$

Solve equations [1] and [2]: $T_1 = 102$ N
$T_2 = 116$ N   (to the nearest 1 N)

In practice you often omit the vector equation and write down the two component equations straight away.
You must say where these equations came from, so your solution to Example 2 would look like this:

Resolving horizontally $\Rightarrow -T_1\cos 27° + T_2\cos 38° = 0$
$T_1\cos 27° = T_2\cos 38°$
Resolving vertically $\Rightarrow T_1\sin 27° + T_2\sin 38° - 12g = 0$
$T_1\sin 27° + T_2\sin 38° = 12g$

In some situations choosing the **i**- and **j**-directions carefully can make the solution easier.

MI

**EXAMPLE 3**

Calculate $T_1$ and $T_2$ in the situation shown.

Take **i** and **j** to be parallel to $BC$ and $BA$.

Resolve in the **i**-direction:

$$T_2 - 12g\cos 32° = 0$$
giving $\qquad T_2 = 99.7\text{ N}$

Resolve in the **j**-direction:

$$T_1 - 12g\sin 32° = 0$$
giving $\qquad T_1 = 62.3\text{ N}$

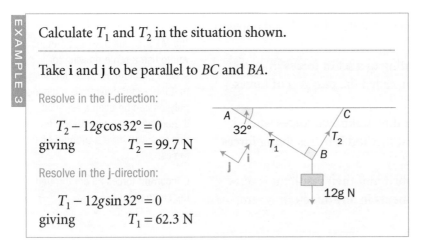

You could alternatively make use of the triangle of forces.

**EXAMPLE 4**

Redo Example 3 using a triangle of forces.

The forces are represented by the sides of the triangle shown.

By simple trigonometry, you have

$$T_1 = 12g\sin 32° = 62.3\text{ N}$$
$$T_2 = 12g\cos 32° = 99.7\text{ N}$$

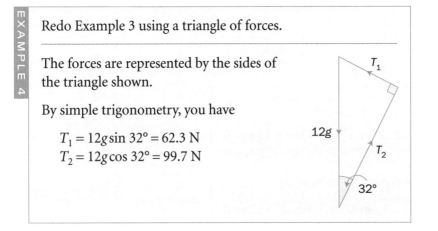

If you have studied the sine rule, all problems involving three forces in equilibrium can be solved using a triangle of forces.

**EXAMPLE 5**

Redo Example 2 using a triangle of forces.

The forces are represented by the sides of the triangle shown.

By the sine rule, you have

$$\frac{T_1}{\sin 52°} = \frac{T_2}{\sin 63°} = \frac{12g}{\sin 65°}$$

which gives

$$T_1 = \frac{12g\sin 52°}{\sin 65°} = 102\text{ N}$$

$$T_2 = \frac{12g\sin 63°}{\sin 65°} = 116\text{ N}$$

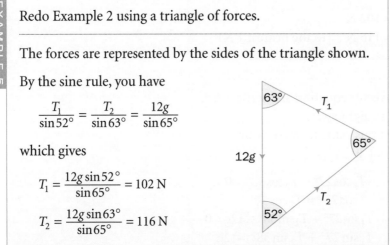

You could write down equations equivalent to those obtained from the sine rule without drawing the triangle using the fact that $\sin\theta = \sin(180° - \theta)$.

This and other trigonometric results are covered in module C2.

$$\frac{T_1}{\sin 52°} = \frac{T_2}{\sin 63°} = \frac{12g}{\sin 65°}$$

is equivalent to

$$\frac{T_1}{\sin 128°} = \frac{T_2}{\sin 117°} = \frac{12g}{115°}$$

These angles are opposite the appropriate forces, as shown.

This rule is known as Lami's theorem.

**Lami's theorem**
For three concurrent forces $P$, $Q$ and $R$ in equilibrium, as shown,

$$\frac{P}{\sin\alpha} = \frac{Q}{\sin\beta} = \frac{R}{\sin\gamma}$$

Knowledge of Lami's theorem is not required by the M1 syllabus, but it will be accepted as a method of solution in the examination.

M1

**EXAMPLE 6**

The diagram shows a block of mass 4 kg resting on a smooth plane inclined at 25° to the horizontal. The block is kept in place by a light string inclined at 15° to the plane. Find the tension, $P$, in the string and the normal reaction, $R$, of the plane on the block.

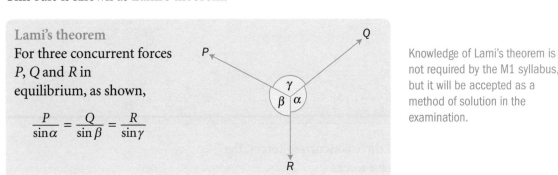

There are three approaches to this problem.

**Resolving forces**
Take the **i**- and **j**-directions parallel and perpendicular to the plane.

Resolve in the **i**-direction:     $P\cos 15° - 4g\sin 25° = 0$     [1]

Resolve in the **j**-direction:     $P\sin 15° + R - 4g\cos 25° = 0$     [2]

This is usually the best choice of **i**- and **j**-directions when dealing with an object on an inclined plane.

Use equation [1] to find P:

$$P = \frac{4g\sin 25°}{\cos 15°} = 17.2\,\text{N}$$

Substitute in equation [2]:

$17.2\sin 15° + R - 4g\cos 25° = 0$   hence $R = 31.1\,\text{N}$

**EXAMPLE 6 (CONT.)**

### Triangle of forces

The three forces are in equilibrium, so the vector diagram is a triangle of forces.

Use the sine rule:

$$\frac{P}{\sin 25°} = \frac{R}{\sin 50°} = \frac{4g}{\sin 105°}$$

$$\Rightarrow \quad P = \frac{4g \sin 25°}{\sin 105°} = 17.2\,\text{N} \quad \text{and} \quad R = \frac{4g \sin 50°}{\sin 105°} = 31.1\,\text{N}$$

### Lami's theorem

Consider the angles between the forces.

Use Lami's theorem:

$$\frac{P}{\sin 155°} = \frac{R}{\sin 130°} = \frac{4g}{\sin 75°}$$

$$\Rightarrow \qquad P = \frac{4g \sin 155°}{\sin 75°} = 17.2\,\text{N}$$

and

$$R = \frac{4g \sin 130°}{\sin 75°} = 31.1\,\text{N}$$

If a problem involves more than three concurrent forces, the recommended method is to resolve forces.

**EXAMPLE 7**

An object $A$ of weight $W$ is suspended by light strings $AB$ and $AC$, of lengths 4 m and 3 m respectively. Points $B$ and $C$ are on the same horizontal level and are 5 m apart. A horizontal force is applied to $A$ so that the tension in $AB$ is twice that in $AC$. Find, in terms of $W$, the value of $P$ and the tension in $AC$.

$ABC$ is a 3-4-5 triangle, so $B\hat{A}C$ is $90°$.

It follows that $\cos\theta = \frac{4}{5}$ and $\sin\theta = \frac{3}{5}$

Consider the forces acting on the object:

Resolving vertically gives $\qquad T\cos\theta + 2T\sin\theta - W = 0$

$$\Rightarrow \qquad \frac{4}{5}T + \frac{6}{5}T = W$$

$$\Rightarrow \qquad T = \frac{1}{2}W$$

Resolving horizontally gives $\quad P + T\sin\theta - 2T\cos\theta = 0$

$$\Rightarrow \qquad P = \frac{8}{5}T - \frac{3}{5}T = T$$

$$\Rightarrow \qquad P = \frac{1}{2}W$$

EXAMPLE 8

A smooth ring, $C$, of mass 3 kg is threaded on a light string 64 cm long. The ends of the string are fixed to points $A$ and $B$, 48 cm apart on the same horizontal level. A force, $P$, is applied to the ring so that it rests vertically below $B$.

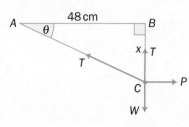

Find the value of $P$ and the tension in the string.

The ring is smooth, so the tension is the same throughout the string.

Let $BC$ be $x$ cm, so that $AC = (64 - x)$ cm.

Use Pythagoras' theorem:

$$x^2 + 48^2 = (64 - x)^2$$

which gives $x = 14$

So, $AC = 50$ cm and $BC = 14$ cm

which give $\sin \theta = \dfrac{7}{25}$

and $\cos \theta = \dfrac{24}{25}$

Consider the forces acting on the ring:

Resolving vertically gives
$T + T \sin \theta - 3g = 0$

$\Rightarrow \qquad \dfrac{32}{25} T = 3g$

$\qquad$ giving $T = 23.0$ N

Resolving horizontally gives
$P - T \cos \theta = 0$

$\Rightarrow \qquad P = \dfrac{24}{25} T$

$\qquad = 22.1$ N

M1

## Exercise 4.2

1 For each of these systems of forces, find the resultant and state the additional force which would be needed to achieve equilibrium.

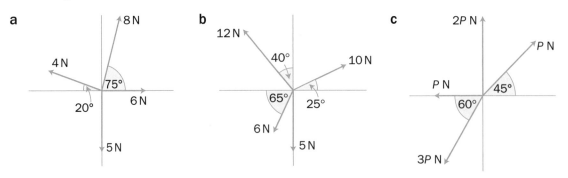

2 Each of these systems of forces is in equilibrium. Find the values of $P$ and $Q$.

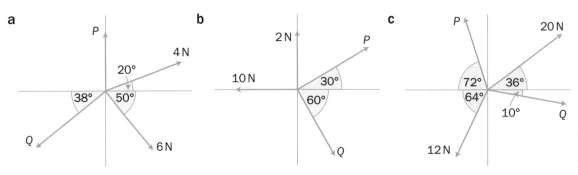

3 Each of these systems of forces is in equilibrium. By drawing a triangle of forces, calculate the unknown quantities.

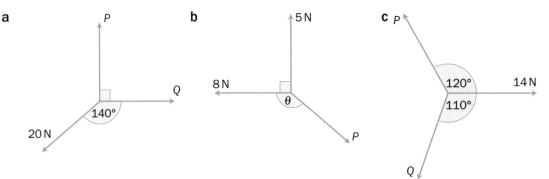

MI

4   Each of these systems of forces is in equilibrium. Use any
    appropriate method to calculate the unknown forces.

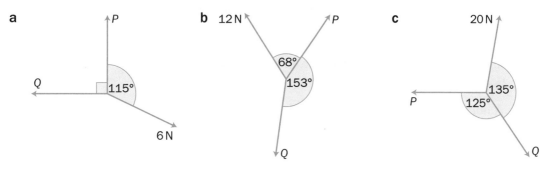

**a**      **b** 12N      **c**

5   A particle of mass 2 kg is suspended by two light strings
    making angles of 30° and 50° with the vertical. Find the
    tensions in the two strings.

6   A block of mass 9 kg rests on a rough plane inclined at 20° to
    the horizontal. Find the magnitudes of the frictional force and
    the normal reaction acting on the block.

7   A block of mass 20 kg rests on smooth horizontal ground near
    to a fixed post, to which it is attached by a light, horizontal rod.
    The block is pulled directly away from the post by a force $P$ N
    inclined at 30° to the horizontal.

    **a**  Find, in terms of $P$, the tension in the rod and the normal
           reaction between the block and the ground.

    **b**  Explain what would happen if the value of $P$ exceeded 392.

8   The diagram shows a cylinder of mass 8 kg lying at rest on
    two smooth planes inclined at angles of 40° and 50° to the
    horizontal.

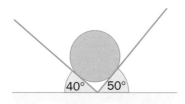

    **a**  Calculate the reaction exerted by each plane on the cylinder.

    **b**  Was it necessary to know that the planes were smooth?

9 A particle, $A$, of weight $W$, is suspended by two strings $AB$ and $AC$. $AB$ is inclined at 30° to the vertical, and $AC$ at an angle $\theta$ to the vertical. The tensions in $AB$ and $AC$ are 40 N and 60 N respectively. Calculate the values of $W$ and $\theta$.

10 A particle of mass 4 kg is suspended from a point $A$ on a vertical wall by a light inextensible string of length 1.3 m.

  a A horizontal force, $P$ N, acting directly away from the wall, is applied to the particle so that it is held in equilibrium at a distance of 0.5 m from the wall.
  Find the value of $P$ and the tension in the string.

  b By drawing a triangle of forces, or otherwise, find the magnitude and direction of the minimum force which would hold the particle in this position.
  Find the tension in the string if this minimum force were applied.

11 Two hooks, $A$ and $B$, are fixed to a ceiling.
  The distance $AB$ is 2.5 m.

  a A particle, $C$, of mass 3 kg is suspended from $A$ and $B$ by two light, inextensible strings. $AC$ is 2 m and $BC$ is 1.3 m. Calculate the tensions in the strings.

  b The two strings are now replaced by a single string of length 3.3 m, which is threaded through a smooth ring attached to the particle. A horizontal force, $P$ N, is applied to the particle so that it rests in the same position as before. Calculate the value of $P$ and the tension in the string.

12 A string is threaded through a smooth ring of weight $W$, and is tied to two points $A$ and $B$ on the same level. The ring is pulled by a force of $P$ N parallel to $AB$, as a result of which it rests in equilibrium with the two parts of the string at angles of 60° and 30° to the vertical.

  a Draw the two possible configurations which fit the given facts.

  b Show that the tension in the string is the same in both cases, and find its value.

  c Find the value of $P$ in each case.

**13** A light string of length $a$ is attached to two points $A$ and $B$ on the same level and a distance $b$ apart, where $b < a$.
A smooth ring of weight $W$ is threaded on the string and is pulled by a horizontal force, $P$, so that it rests in equilibrium vertically below $B$.

**a** Show that the tension in the string is $\dfrac{W(a^2 + b^2)}{2a^2}$

**b** Find the force $P$.

**14** A particle of weight $W$ is attached by a light inextensible string of length $a$ to a point $A$ on a vertical wall. The particle is supported in equilibrium by a light, rigid rod of length $b$ attached to a point $B$ on the wall at a distance $a$ vertically below $A$.

**a** Show that the tension in the string is $W$.

**b** Find the thrust in the rod.

**15** If an object is suspended from a single point, the centre of gravity of the object must be directly below the point of suspension.

A uniform beam $AB$, of length 5.2 m and mass 50 kg, is suspended in equilibrium by two light inextensible strings $OA$ and $OB$ attached to the ends of the beam and to a fixed point $O$. The strings have lengths 4.8 m and 2 m respectively.

**a** Draw a diagram to show this situation.

**b** Show that the tension in $OA$ is approximately 452 N, and calculate the tension in $OB$.

M1

## 4.3 Modelling friction

You were introduced to the concept of a frictional force in 4.1.
In this section you will explore how such forces can be modelled.

Your experience of friction should support the following facts.

- Friction occurs when one surface slides over another surface which is in contact with it.
- Friction always acts in the direction opposite to any movement or tendency to move.
- In any situation there is a maximum possible frictional force. If the applied force is greater than this maximum, movement will take place.

> You should ideally do some practical investigations to get a feel for the way friction behaves.

Consider a block, of weight $W$, placed on a horizontal surface.
A horizontal force, $P$, is applied to the block.

### INVESTIGATION

Provided the block is in equilibrium,
$$R = W$$
and $F = P$

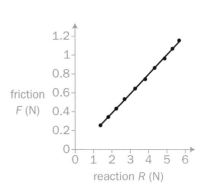

Increase $P$ until the limit is reached and the block is about to start moving.

At this point the value of $P$ gives the maximum possible frictional force.

Here are some practical results obtained for various values of $W$:

| Normal reaction $R$ (N) | Maximum friction $F$ (N) | $\frac{F}{R}$ |
|---|---|---|
| 1.37 | 0.294 | 0.214 |
| 1.86 | 0.392 | 0.211 |
| 2.35 | 0.490 | 0.208 |
| 2.84 | 0.588 | 0.207 |
| 3.43 | 0.686 | 0.200 |
| 4.02 | 0.784 | 0.195 |
| 4.51 | 0.882 | 0.196 |
| 5.00 | 0.980 | 0.196 |
| 5.49 | 1.08 | 0.196 |
| 5.88 | 1.18 | 0.200 |

The ratio between $F$ and $R$ is constant (allowing for slight experimental error).

Experiments such as this lead to the following model.

- Friction increases to match the applied force up to a maximum, the limiting friction. The system is then in limiting equilibrium.
- The limiting frictional force depends only on the type of surfaces and the normal reaction between them.
- For a given pair of surfaces, the ratio between the limiting friction and the normal reaction is constant.

In limiting equilibrium even a slight increase in the applied force would cause the system to move.

The frictional force is independent of the area of contact between the surfaces.

The ratio between limiting friction and normal reaction is the coefficient of friction, denoted by $\mu$. This ratio has different values for different surfaces.

As the frictional force is always less than or equal to the limiting friction,

$$\frac{F}{R} \leqslant \mu \quad \text{or} \quad F \leqslant \mu R$$

where $\mu$ is the coefficient of friction, which is constant for a given pair of surfaces.

In reality, the frictional force when the object is moving (the dynamic friction) is usually slightly less than the maximum friction when it is stationary (the static friction).
However, the M1 module assumes they are the same, so

when a body is moving or on the point of moving

$$F = \mu R$$

EXAMPLE 1

A block of mass 4 kg rests on a rough horizontal surface. The coefficient of friction between the block and the surface is 0.35. A horizontal force, $P$, is applied to the block so that it is just on the point of moving. Find the value of $P$.

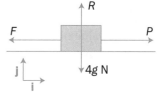

Resolve in the i-direction:
$$P - F = 0 \quad \Rightarrow \quad P = F \qquad [1]$$

Resolve in the j-direction:
$$R - 4g = 0 \quad \Rightarrow \quad R = 4g \qquad [2]$$

As friction is limiting,

$$\frac{F}{R} = 0.35 \quad \Rightarrow \quad F = 0.35R \qquad [3]$$

$\mu = 0.35$

Substitute from [2] into [3]:
$$F = 0.35 \times 4g = 13.7 \text{ N}$$
Now use equation [1]: $\quad P = F = 13.7 \text{ N}$

**EXAMPLE 2**

Repeat Example 1, but with the applied force, $P$, inclined at 20° above the horizontal.

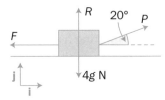

Resolve in the i-direction:
$$P \cos 20° - F = 0 \qquad [1]$$
Resolve in the j-direction:
$$R + P \sin 20° - 4g = 0 \qquad [2]$$
As friction is limiting,
$$F = 0.35R \qquad [3]$$

Use [2] and [3] to find $F$:
$$F = 0.35(4g - P \sin 20°)$$
Substitute in equation [1]:
$$P \cos 20° = 0.35(4g - P \sin 20°)$$

$$\Rightarrow \qquad P = \frac{0.35 \times 4g}{\cos 20° + 0.35 \sin 20°}$$
$$= 13.0 \text{ N}$$

**EXAMPLE 3**

The diagram shows a block of mass 5 kg resting on a rough plane inclined at 30° to the horizontal. A horizontal force of 70 N is applied and the block is on the point of moving up the slope.

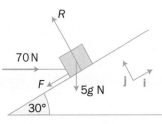

Find the value of the coefficient of friction between the block and the plane.

Resolve in the i-direction (up the slope):
$$70 \cos 30° - F - 5g \sin 30° = 0 \qquad [1]$$
Resolve in the j-direction (perpendicular to the slope):
$$R - 70 \sin 30° - 5g \cos 30° = 0 \qquad [2]$$
Use equation [1] to find $F$: $\qquad F = 36.1 \text{ N}$
Now use equation [2] to find $R$: $\qquad R = 77.4 \text{ N}$
As the block is about to move, $F = \mu R$.
Hence $36.1 = 77.4\mu$
$$\Rightarrow \qquad \mu = 0.466$$

In problems involving inclined planes it is usually best to resolve parallel and perpendicular to the slope rather than horizontally and vertically.

EXAMPLE 4

A block of weight $W$ rests on a rough plane inclined at an angle $\alpha$ to the horizontal. The value of $\alpha$ is increased until the block is on the point of slipping. Show that the coefficient of friction $\mu = \tan\alpha$.

This is a useful fact to remember, and provides a simple way of finding an experimental value for the coefficient of friction.

Draw a triangle of forces for this situation:

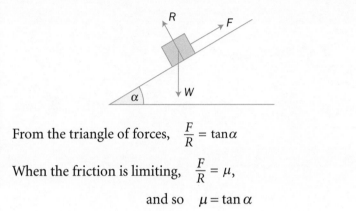

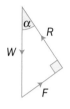

From the triangle of forces, $\quad \dfrac{F}{R} = \tan\alpha$

When the friction is limiting, $\quad \dfrac{F}{R} = \mu,$

and so $\quad \mu = \tan\alpha$

## Exercise 4.3

1 Each of these diagrams shows a block of mass 5 kg resting on a rough horizontal surface. The block is in limiting equilibrium. Find the coefficient of friction.

a

b

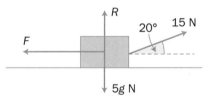

c

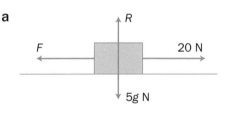

d

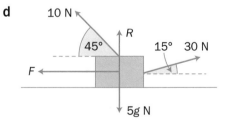

e

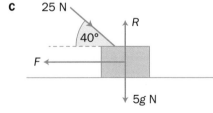

f

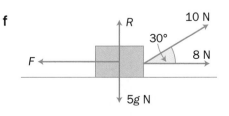

2 Each of these diagrams shows a block of mass 8 kg resting on
a rough horizontal surface. The block is in limiting equilibrium.
Find the force, $P$, for the stated value of $\mu$.

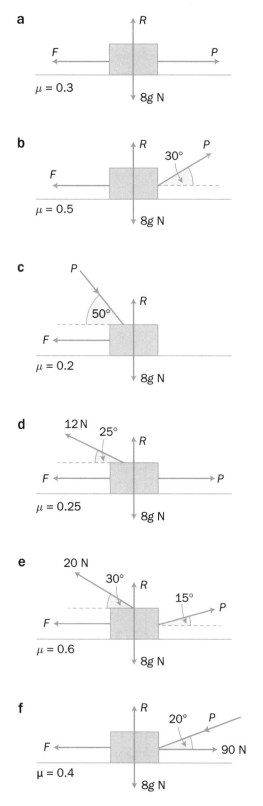

**a**

$\mu = 0.3$

**b**

$\mu = 0.5$

**c**

$\mu = 0.2$

**d**

12 N  25°

$\mu = 0.25$

**e**

20 N  30°  15°

$\mu = 0.6$

**f**

20°

90 N

$\mu = 0.4$

MI

3 A block of mass 5 kg rests in limiting equilibrium on a rough plane inclined at 27° to the horizontal. Find the magnitude of the friction force acting on the block and the coefficient of friction between the block and the plane.

4 A block of mass 4 kg rests on a rough plane inclined at 10° to the horizontal. The coefficient of friction between the block and the plane is 0.3. A force $P$ acts on the block parallel to the plane. Find the magnitude and direction of $P$ if the block is about to move

**a** up the plane

**b** down the plane.

5 A block of mass 4 kg is on a rough plane inclined at 30° to the horizontal, as shown, with a horizontal force $P$ acting on it.

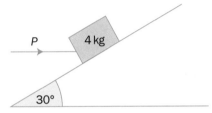

The coefficient of friction between the block and the plane is 0.4. Find the range of possible values of $P$ if the block is to remain stationary.

6 The diagram shows a 6 kg block resting on a rough horizontal table. It is connected by light, inextensible strings passing over two smooth pulleys to two blocks of masses 2 kg and 5 kg which hang vertically.

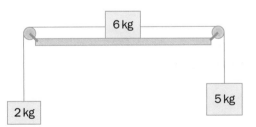

Calculate the coefficient of friction between the 6 kg block and the table if the system is on the point of moving.

7 The diagram shows a block of mass 3 kg resting on a smooth plane inclined at 40° to the horizontal. It is connected by a light string passing over a smooth pulley to a second block, of mass 4 kg, which is resting on a rough horizontal surface.

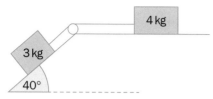

The system is on the point of moving.

**a** By considering the forces acting on the 3 kg block, find the tension in the string.

**b** By considering the forces acting on the 4 kg block, find the coefficient of friction between it and the horizontal surface.

8 A particle of mass $m$ can just rest on a rough plane inclined at 30° to the horizontal without slipping down. The inclination of the plane is then increased to 45°, and the particle is kept at rest by a horizontal force $P$. Show that the least possible value of $P$ is $0.268mg$.

9 The diagram shows blocks of mass 10 kg and 4 kg resting on two rough planes inclined at 40° to the horizontal.
The blocks are connected by a light string passing over a smooth pulley at the top of the planes.

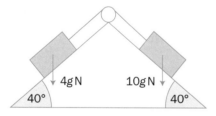

The coefficient of friction for each block is $\mu$ and the system is in limiting equilibrium. Find the value of $\mu$.

10 The diagram shows blocks of masses 2 kg and 1 kg resting on two rough planes inclined to the horizontal at 60° and 30° respectively. The blocks are connected by a light string passing over a smooth pulley at the top of the planes.

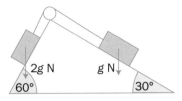

The coefficient of friction between each block and plane is $\mu$. Show that if the system is on the point of slipping then $\mu = 0.660$.

11 A block of mass $m$ can rest without slipping on a rough plane inclined at an angle $\alpha$ to the horizontal. If a force $P_1$ acting directly up the plane is applied to the block, the block is on the point of moving in that direction. If a force $P_2$ acting directly down the plane is applied to the block, the block is on the point of moving in that direction. Show that the magnitudes of $P_1$ and $P_2$ differ by an amount which is independent of the coefficient of friction between the block and the plane.

12 A particle of weight $W$ N rests on a rough plane inclined at an angle $\alpha$ to the horizontal. The coefficient of friction between the particle and the plane is $\mu$. A force of $P$ N parallel to the plane is just sufficient to prevent the particle from sliding down the plane. If a force of $kP$ N is applied parallel to the plane, the particle is on the point of moving up the plane. Find an expression for $k$ in terms of $\mu$ and $\alpha$.

Unless stated otherwise you can let $g = 9.8 \, \mathrm{m s}^{-2}$

**1** The diagram shows an object of mass 50 kg supported by two
light strings, one horizontal and the other inclined at 30° to
the vertical. The tensions in the strings are $T_1$ and $T_2$, as shown.

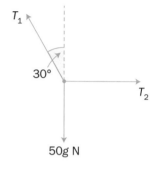

**a** Show that $T_1 = 566$ N.

**b** Find the value of $T_2$.

**2** An object of mass 20 kg is supported in equilibrium by two
light, inextensible strings inclined at angles of 35° and 55°
to the vertical.
Find the tensions in the strings.

**3** The diagram shows a particle of mass 30 kg resting in equilibrium
on a smooth plane inclined at 25° to the horizontal. It is supported
in this position by a light, inextensible string inclined at 35°
to the plane, as shown.

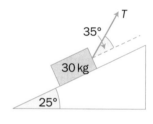

**a** Show that the tension, $T$, in the string is approximately 152 N.

**b** Find the normal reaction between the particle and the plane.

**4** A particle $P$ of weight 6 N is attached to one end of a light
inextensible string. The other end of the string is attached to
a fixed point $O$. A horizontal force of magnitude $F$ newtons is
applied to $P$. The particle $P$ is in equilibrium under gravity
with the string making an angle of 30° with the vertical,
as shown in the diagram.
Find, to three significant figures

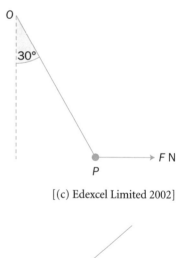

**a** the tension in the string

**b** the value of $F$.

[(c) Edexcel Limited 2002]

**5** Two forces **P** and **Q**, act on a particle. The force **P** has
magnitude 5 N and the force **Q** has magnitude 3 N.
The angle between the directions of **P** and **Q** is 40°, as
shown in the diagram. The resultant of **P** and **Q** is **F**.

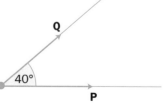

**a** Find, to three significant figures, the magnitude of **F**.

**b** Find, in degrees to one decimal place, the angle between
the directions of **F** and **P**.

[(c) Edexcel Limited 2001]

6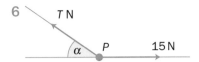

A particle $P$ of mass 2 kg is held in equilibrium under gravity by two light inextensible strings. One string is horizontal and the other is inclined at an angle $\alpha$ to the horizontal, as shown in the diagram. The tension in the horizontal string is 15 N. The tension in the other string is $T$ N.
By drawing a triangle of forces, or otherwise:

**a**  Find the size of the angle $\alpha$.

**b**  Find the value of $T$.

7  A box of mass 50 kg rests on a rough horizontal floor. The coefficient of friction between the box and the floor is $\mu$. A force of 100 N is applied to the box by means of a rope inclined at 30° to the horizontal, as shown.

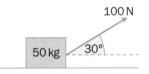

**a**  Show that the normal reaction between the floor and the box is 440 N.

**b**  If the box is in limiting equilibrium, find the value of $\mu$.

8  A ring of mass 0.3 kg is threaded on a fixed, rough horizontal curtain pole. A light inextensible string is attached to the ring. The string and the pole lie in the same vertical plane. The ring is pulled downwards by the string which makes an angle $\alpha$ to the horizontal, where $\tan \alpha = \dfrac{3}{4}$ as shown in the diagram.

Use $g = 10 \text{ m s}^{-2}$ in this question.

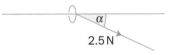

The tension in the string is 2.5 N. Given that, in this position, the ring is in limiting equilibrium,

**a**  find the coefficient of friction between the ring and the pole.

The direction of the string is now altered so that the ring is pulled upwards. The string lies in the same vertical plane as before and again makes an angle $\alpha$ with the horizontal, as shown in the diagram. The tension in the string is again 2.5 N.

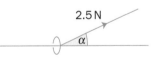

**b**  Find the normal reaction exerted by the pole on the ring.

**c**  State whether the ring is in equilibrium in the second position, giving a brief justification for your answer. You need make no further detailed calculation of the forces acting.

[(c) Edexcel Limited 2002]

M1

## Summary

Refer to

- Mass (in kilograms) is the amount of matter forming an object.
  Weight (in newtons) is the force on an object in a gravitational field.
  - Acceleration of gravity near Earth's surface is $g \approx 9.8\ \mathrm{m\,s^{-2}}$
  - For an object of mass $m$, weight = $mg$ 4.1
- In addition to weight you meet four main types of force: 4.1

| Type of force | Where? | How caused? | Direction |
|---|---|---|---|
| Tension | In a string or rod | when both ends are pulled | along string/rod |
| Thrust (compression) | In a rod | when both ends are pushed | along rod |
| Normal reaction | Between surfaces in contact | when the surfaces are pressed together | perpendicular to plane of contact |
| Friction | Between surfaces in contact | when one surface slides over the other | parallel to plane of contact and opposing motion |

- Forces that are concurrent (pass through the same point) are in equilibrium if
  - the sum of their components in any direction is zero
  - a vector diagram representing the forces forms a closed polygon. 4.2
- To find unknown forces which are in equilibrium, you can
  - form equations by resolving in two perpendicular directions
  - draw and solve a triangle of forces for three-force problems
  - for three-force problems, use Lami's theorem. 4.2
- The standard model of friction is that
  - the frictional force acts to oppose potential or actual motion
  - if the object is moving or on the point of moving 4.3
    (limiting equilibrium), the frictional force is given by $F = \mu R$,
    where $R$ is the normal reaction and $\mu$ is the coefficient of friction.

---

### Links

Police use the analysis of frictional forces as part of the
investigation of a road accident.
Before the collision the driver brakes suddenly causing tyre
marks on the road. Police can use the tyre marks and information
about the coefficient of friction to form equations. Solving these
equations allows them to determine the speed of the vehicle
before the incident occurred.

M1

1 A smooth plane is inclined at an angle of 20° to the horizontal.
A particle $P$ of mass 3 kg is held in equilibrium on the plane
by a horizontal force of magnitude $F$ newtons, as shown
in the diagram.
Find, to three significant figures,

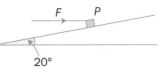

a the normal reaction exerted by the plane on $P$

b the value of $F$.

2 Hari and Ian are coming to the end of a race. Hari is 10 m
ahead of Ian and is running at a constant 8 m s$^{-1}$. Ian has 50 m
to go to the finishing line. He is running at 7 m s$^{-1}$, but
accelerating at a uniform rate, so that the result of the race is a
dead heat. Calculate

a the speed Ian is running at as he crosses the finishing line

b the acceleration Ian maintained over the final 50 m.

3 The diagram shows forces of 5 N and 8 N acting so that their
resultant, $R$ N, is in the **i**-direction. The forces act at angles
of $\theta$ and 30° to the resultant, as shown.

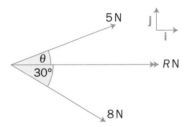

a Express the two forces in component form relative to the
**i**- and **j**-directions shown.

b Hence calculate
i the angle $\theta$
ii the value of $R$.

4 The diagram shows a block of mass 10 kg resting on a rough
plane inclined at 20° to the horizontal. A force of magnitude
$P$ N acts parallel to the plane, as shown. When $P = 30$, the
block is on the point of slipping down the slope.

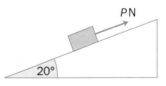

a Calculate the coefficient of friction between the block and
the plane.

The force $P$ N is increased until the block is on the point of
sliding up the slope.

b Find the new value of $P$.

M1

5 Annette, Barbara and Casey are arguing over possession of a toy.
They each grab hold of it and pull horizontally. Annette exerts
a force of 50 N, Barbara a force of 100 N and Casey 150 N.
The angle between the directions of each pair of forces is 120°.

   **a** Calculate the magnitude of the resultant force exerted on
the toy.

   **b** Calculate the angle between the direction of the resultant force
and the direction in which Casey is pulling.

6 A post office van sets off from rest after making a delivery. It accelerates
at a constant rate of 1.5 m s$^{-2}$ until reaching a speed of 12 m s$^{-1}$, and then
immediately decelerates at 3 m s$^{-2}$ until it stops for the next delivery.

   **a** Sketch a speed–time graph to illustrate this journey.

   **b** Hence, or otherwise, find the distance between the two
delivery points.

7

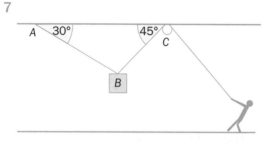

The diagram shows a box, *B*, of mass 100 kg, attached by a light rope *AB*
to a fixed point on the ceiling. A second similar rope is attached to *B* and
passes over a smooth pulley at *C*, on the same level as *A*. George hauls on
the rope, as shown, and raises the box so that *AB* and *BC* make angles
of 30° and 45° respectively with the horizontal.

   **a** Draw a diagram showing the forces acting on the box.

   **b** Calculate the tensions in the ropes *AB* and *BC*.

   **c** Find the force which George is exerting on the rope.

The rope has been described as 'light' and the pulley as 'smooth'.

   **d** Explain what these two modelling assumptions allow you
to assume about the tensions in the ropes.

George now moves so that the section of rope he is holding is
vertical, without changing the position of the box.

   **e** Without making detailed calculations, explain what effect
this change has on

     **i** the magnitude of the force George exerts on the rope
     **ii** the magnitude of the total force acting on the pulley.

M1

**8** A ball is projected vertically upwards from ground level. It returns to ground level after 6 s. Assuming that air resistance can be ignored, calculate

   **a** the height to which the ball travels

   **b** the speed at which it was projected

   **c** the speed at which it returned to ground level

   **d** the length of time for which the ball was over 20 m above the ground.

In fact the ball was sufficiently large that air resistance had a significant effect.

   **e** Assuming the initial speed was the value found in part **b**, explain the effect of air resistance on
     **i** the maximum height reached by the ball
     **ii** the speed at which it returned to ground level.

**9** A particle has mass 2 kg. It is attached at $B$ to the ends of two light inextensible strings $AB$ and $BC$. When the particle hangs in equilibrium, $AB$ makes an angle of 30° with the vertical, as shown. The magnitude of the tension in $BC$ is twice the magnitude of the tension in $AB$.

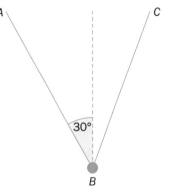

   **a** Find, in degrees to one decimal place, the size of the angle that $BC$ makes with the vertical.

   **b** Hence find, to three significant figures, the magnitude of the tension in $AB$.

[(c) Edexcel Limited 2002]

**10** A train starts from rest at station $A$ and moves along a straight horizontal track. For the first 10 s, the train moves with constant acceleration 1.2 m s$^{-2}$. For the next 24 s it moves with constant acceleration 0.75 m s$^{-2}$. It then moves with constant speed for $T$ seconds. Finally it slows down with constant deceleration 3 m s$^{-2}$ until it comes to rest at station $B$.

   **a** Show that, 34 s after leaving $A$, the speed of the train is 30 m s$^{-1}$.

   **b** Sketch a speed–time graph to illustrate the motion of the train as it moves from $A$ to $B$.

   **c** Find the distance moved by the train during the first 34 s of its journey from $A$.

The distance from $A$ to $B$ is 3 km.

   **d** Find the value of $T$.

[(c) Edexcel Limited 2004]

M1

**11** A box of mass 1.5 kg is placed on a plane which is inclined at an angle of 30° to the horizontal. The coefficient of friction between the box and plane is $\frac{1}{3}$. The box is kept in equilibrium by a light string which lies in a vertical plane containing a line of greatest slope of the plane. The string makes an angle of 20° with the plane, as shown in the diagram. The box is in limiting equilibrium and is about to move up the plane. The tension in the string is $T$ newtons. The box is modelled as a particle. Find the value of $T$.

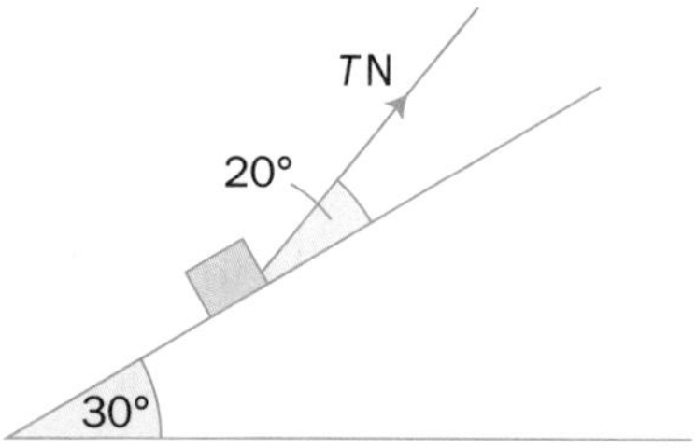

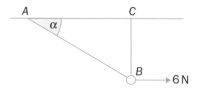

[(c) Edexcel Limited 2003]

**12** A smooth bead, $B$, is threaded on a light inextensible string. The ends of the string are attached to two fixed points $A$ and $C$ on the same horizontal level. The bead is held in equilibrium by a horizontal force of magnitude 6 N acting parallel to $AC$. The bead $B$ is vertically below $C$ and $\angle BAC = \alpha$, as shown in the diagram. Given that $\tan \alpha = \frac{3}{4}$, find

**a**   the tension in the string

**b**   the weight of the bead.

[(c) Edexcel Limited 2005]

**13** Two ships, $P$ and $Q$, are travelling at night with constant velocities. At midnight, $P$ is at the point with position vector $(20\mathbf{i} + 10\mathbf{j})$ km relative to a fixed origin $O$. At the same time, $Q$ is at the point with position vector $(14\mathbf{i} - 6\mathbf{j})$ km. Three hours later, $P$ is at the point with position vector $(29\mathbf{i} + 34\mathbf{j})$ km. The ship $Q$ travels with velocity $12\mathbf{j}$ km h$^{-1}$. At time $t$ hours after midnight, the position vectors of $P$ and $Q$ are $\mathbf{p}$ km and $\mathbf{q}$ km respectively. Find

**a**   the velocity of $P$, in terms of $\mathbf{i}$ and $\mathbf{j}$

**b**   expressions for $\mathbf{p}$ and $\mathbf{q}$, in terms of $t$, $\mathbf{i}$ and $\mathbf{j}$.

At time $t$ hours after midnight, the distance between $P$ and $Q$ is $d$ km.

**c**   By finding an expression for $\overrightarrow{PQ}$, show that
$d^2 = 25t^2 - 92t + 292$.

Weather conditions are such that an observer on $P$ can only see the lights on $Q$ when the distance between $P$ and $Q$ is 15 km or less. Given that when $t = 1$, the lights on $Q$ move into sight of the observer,

**d**   find the time, to the nearest minute, at which the lights on $Q$ move out of sight of the observer.

[(c) Edexcel Limited 2005]

# 5

# Newton's laws of motion

This chapter will show you how to
- model the relationship between force and acceleration using Newton's laws
- write down the equation of motion for an object acted on by forces
- solve problems involving connected particles.

## Before you start

### You should know how to:

**3.2** **1** Resolve a force into components.

**4.3** **2** Use the standard model of friction.

**2.4** **3** Use the formulae relating to motion with constant acceleration.

### Check in

**1**

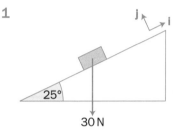

The diagram shows a block of weight 30 N resting on a slope inclined at 25° to the horizontal. Express the weight as a vector in component form relative to the **i**- and **j**-directions indicated.

**2** A block of mass 5 kg rests on a horizontal surface. The coefficient of friction between the block and the surface is 0.4.
A horizontal force, $P$ N, is applied and the block is on the point of moving.
Find the value of $P$.

**3** A car moving in a straight line at 10 m s$^{-1}$ accelerates to a speed of 20 m s$^{-1}$ while travelling 100 m. Find

**a** its acceleration

**b** the length of time for which it was accelerating.

Newton's laws model the relationship between the motion of an object and the forces acting on it.

Another model — Einstein's theory of relativity — gives predicted results which differ from those of Newton only at the atomic or astronomical scale, where Einstein's model is superior. For most practical purposes you still use Newton's model for analysing the motion of systems.

Newton's first law shows that force is related to acceleration.

**Newton's first law of motion**
**Every object remains at rest or moves with constant velocity unless an external force is applied.**

Of course, if there are several forces acting on the object, it may be in equilibrium, in which case there would be no acceleration. The object will only speed up, slow down or change direction if the forces combine to a give a non-zero **resultant force**.

In other words, if you want an object to accelerate, you have to apply a force.

The relationship between the force and the acceleration is covered by the second law.

**Newton's second law of motion**
**When an applied force causes an object to accelerate**
○ **the force and acceleration are in the same direction**
○ **the magnitude of the force is proportional to the magnitude of the acceleration and to the mass of the object.**

Newton actually stated the second law in terms of momentum, but for your purposes it is better to talk in terms of mass and acceleration.

This law fits well with common sense.
○ For a given object, a larger acceleration needs a larger force.
○ The more massive the object, the greater the force needed to produce a given acceleration.

Symbolically, Newton's second law is

$$\mathbf{F} \propto m\mathbf{a} \quad \text{or} \quad \mathbf{F} = km\mathbf{a} \quad \text{where } k \text{ is a constant.}$$

This is a relation between **vectors** since the force and acceleration have the same direction.

The SI unit of force is the newton, which is defined as the force needed to accelerate a 1 kg mass at 1 m s$^{-2}$. With this definition the value of $k$ is 1, and Newton's second law becomes

$$\mathbf{F} = m\mathbf{a}$$

This is the equation of motion of the object.
In words, you can say
    'Resultant force equals mass times acceleration.'

It is important to remember that **F** is the **resultant** of the forces acting on the object.

MI

EXAMPLE 1

The engine of a car of mass 900 kg produces a driving force of 2000 N. There are resistive forces of 650 N. Find the acceleration of the car on a level road.

Draw a diagram to show the forces acting:

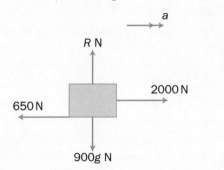

Acceleration takes place in a horizontal direction, so the only forces to consider are those acting horizontally.

Resolve horizontally:

Resultant force = 2000 − 650 = 1350 N

Use the equation of motion $F = ma$:

$1350 = 900a$

$\Rightarrow \quad a = 1.5$ m s$^{-2}$ horizontally.

The equation of motion is a vector equation, and would formally be written as
$1350\mathbf{i} = 900\mathbf{a} \Rightarrow \mathbf{a} = 1.5\mathbf{i}$ m s$^{-2}$
However, when motion is in one dimension you can write the equation in scalar form, as in this example.

EXAMPLE 2

A horizontal force of 50 N is applied to a sledge of mass 20 kg resting on level snow. The sledge accelerates at 2.2 m s$^{-2}$. Find the coefficient of friction between the sledge and the snow.

Draw a diagram:

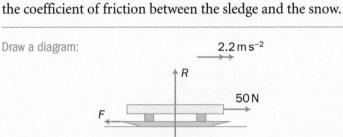

Resolve horizontally and apply Newton's second law:

$50 - F = 20 \times 2.2$

$\Rightarrow \quad F = 6$ N

$F$ is the frictional force.

Resolve vertically:

$R - 20g = 0$

$R = 20g = 196$ N

There is no vertical acceleration, so the resultant of the vertical forces is zero.

The sledge is moving, so

$$\frac{F}{R} = \mu$$

$\Rightarrow \quad \mu = \dfrac{6}{196} = 0.0306$

In the next example you need to use vectors.

EXAMPLE 3

An object of mass 10 kg is acted on by forces $(3\mathbf{i} + 6\mathbf{j})$ N, $(2\mathbf{i} - 3\mathbf{j})$ N and $(\mathbf{i} + 2\mathbf{j})$ N, relative to some coordinate system. Find the acceleration of the object.

The resultant force acting on the object is

$$(3\mathbf{i} + 6\mathbf{j}) + (2\mathbf{i} - 3\mathbf{j}) + (\mathbf{i} + 2\mathbf{j}) = 6\mathbf{i} + 5\mathbf{j}$$

Let the acceleration be **a**.

Apply Newton's second law:

$$6\mathbf{i} + 5\mathbf{j} = 10\mathbf{a}$$
$$\mathbf{a} = 0.6\mathbf{i} + 0.5\mathbf{j}$$

This is the acceleration in component form.

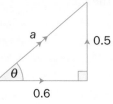

You could then, if required, find its magnitude and direction:

$$|\mathbf{a}| = \sqrt{0.6^2 + 0.5^2} = 0.781 \text{ m s}^{-2}$$

If $\theta$ is the angle with the **i**-direction, then

$$\tan\theta = \frac{0.5}{0.6}$$
$$\Rightarrow \quad \theta = 39.8°$$

## Exercise 5.1

1  A body of mass 40 kg is acted upon by a resultant force of 90 N. Find the acceleration of the body.

2  Find the force needed to accelerate a body of mass 25 kg at 2.1 m s$^{-2}$.

3  A body is acted upon by a force of 24 N and undergoes an acceleration of 3.6 m s$^{-2}$. What is the mass of the body?

4  A body of mass 4 kg is acted upon by a resultant force of $(12\mathbf{i} + 18\mathbf{j})$ N. Find, in component form, the acceleration of the body.

5  A particle of mass 3 kg undergoes an acceleration of $(2\mathbf{i} - 5\mathbf{j})$ m s$^{-2}$. What is the resultant force acting on the body?

6 The table shows information about a vehicle moving on a level road. Find the missing quantities.

|   | Driving force (N) | Resistance (N) | Mass (kg) | Acceleration (m s$^{-2}$) |
|---|---|---|---|---|
| a | 1200 | 800 | 500 | |
| b | 2000 | 600 | | 3.5 |
| c | 900 | | 650 | 0.8 |
| d | | 250 | 800 | 1.3 |
| e | 500 | 800 | 750 | |

7 Find, in component form, the acceleration of a body of mass 4 kg acted upon by forces $(5\mathbf{i} + \mathbf{j})$ N, $(2\mathbf{i} + 7\mathbf{j})$ N and $(-4\mathbf{i} - 3\mathbf{j})$ N.

8 A body of mass 2 kg is acted upon by forces $(2\mathbf{i} + 4\mathbf{j})$ N, $(3\mathbf{i} - 5\mathbf{j})$ N and an unknown force **P**. Find the force **P**, given that the acceleration of the body is $(2\mathbf{i} - \mathbf{j})$ m s$^{-2}$.

9 A block of mass 3 kg is being towed across a horizontal surface, with coefficient of friction 0.2, by a horizontal force of 18 N. Find the acceleration of the block.

10 A block of mass 5 kg is being towed across a horizontal surface, with coefficient of friction $\mu$, by a horizontal force of 40 N. If the acceleration of the block is 5 m s$^{-2}$, find the value of $\mu$.

11 A horse is towing a truck along rails. The horse is attached to the truck by a rope of negligible mass which is horizontal and makes an angle of 20° with the direction of the rails. The truck has a mass of 1200 kg and its motion is opposed by a resistance force of 300 N. Find the tension in the rope if the acceleration of the truck is 0.3 m s$^{-2}$.

12 Find the magnitude and direction of the acceleration of each of these objects.

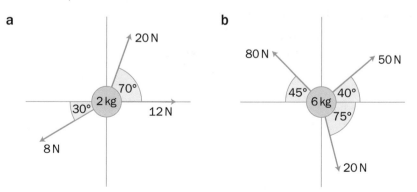

a

b

M1

13 An object of mass 12 kg is pulled up a smooth slope, inclined at 45° to the horizontal, by a string parallel to the slope.

    a If the tension in the string is 120 N, find the acceleration of the object.

    b If the tension is then reduced so that the object has an acceleration down the slope of 2 m s$^{-2}$, find the new tension.

14 The diagram shows a truck of mass 500 kg moving along rails under the action of a force $P$ N applied at 20° to the direction of the rails.

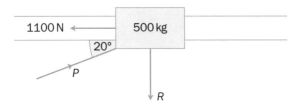

There are resistive forces totalling 1100 N, as shown. $R$ is the horizontal normal reaction of the rails on the wheels of the truck. For safety reasons, $R$ should not exceed 500 N.

    a Find the maximum safe value of $P$.

    b Find the greatest safe acceleration of the truck.

15 Rory, Anna and Dave are three lions fighting over a piece of meat of mass 12 kg. Each lion exerts a horizontal pull. Rory pulls with a force of 800 N. Anna, who is 120° anticlockwise from Rory, exerts a force of 400 N. Dave is 140° clockwise from Rory. The meat accelerates in Rory's direction.

    a Find the force which Dave is exerting.

    b Find the magnitude of the acceleration.

16 A boat of mass 3 tonnes is steered due east with its engines exerting a driving force of 4000 N. A wind blowing from the south exerts a force of 1200 N. There is a resistance of 2000 N opposing motion. Find the magnitude and direction of the boat's acceleration.

17

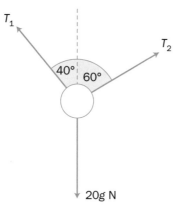

$T_1$  40°  60°  $T_2$

20g N

The diagram shows a particle of mass 20 kg suspended in equilibrium by strings making angles of 40° and 60° with the vertical.

**a** Calculate the tensions in the strings.

The string with the greatest tension suddenly breaks.

**b** Find the magnitude and direction of the initial acceleration of the particle.

18 Irene gets on her bicycle at the top of an incline, which is 200 m long and at an angle of 10° to the horizontal. She freewheels down the slope from rest. The total mass of Irene and her bicycle is 100 kg and there is a constant resistance of 40 N.

**a** Find the speed with which Irene reaches the bottom of the slope.

The road then slopes upwards at an angle $\theta$ to the horizontal. Irene freewheels up this slope and travels 80 m before coming to rest.

**b** Assuming that the resistance remained constant, find the value of $\theta$.

M1

You will now see the reason for the relation between mass and weight introduced in 4.1.

An object falling freely near the surface of the Earth accelerates at approximately 9.8 m s$^{-2}$.
This is the acceleration due to gravity, and is denoted by $g$.

The value of this acceleration varies slightly depending on where you conduct the experiment.

Because the object is accelerating, there must be a downward force, $W$, acting on it. If the mass is $m$, then by Newton's second law you have

$$W = mg$$

The force $W$ is called the weight of the object.

Mass is constant, but weight depends on the gravitational acceleration.
e.g. An object of mass 10 kg has a weight of $10 \times 9.8 = 98$ N near the Earth's surface. On the Moon, the same object would have a weight of $10 \times 1.6 = 16$ N, because gravitational acceleration on the Moon is approximately 1.6 m s$^{-2}$.

**EXAMPLE 1**

A crane lifts a 120 kg object on the end of its cable, which has negligible mass. At first the object accelerates at 2 m s$^{-2}$. It then travels at a uniform speed and finally it slows to rest with an acceleration of $-1.2$ m s$^{-2}$. Find the tension in the cable during each stage of the motion.

The weight of the object is $120 \times 9.8 = 1176$ N

Consider each stage separately:

**Stage 1**    Resolve vertically and use $F = ma$:
$$T - 1176 = 120 \times 2 \quad \Rightarrow \quad T = 1416 \text{ N}$$

**Stage 2**    There is no acceleration and so no resultant force. The tension and the weight must be equal.

Resolve vertically:
$$T = 1176 \text{ N}$$

**Stage 3**    Resolve vertically and use $F = ma$:
$$T - 1176 = 120 \times (-1.2) \quad \Rightarrow \quad T = 1032 \text{ N}$$

EXAMPLE 2

An object of mass 8 kg is being towed by a light string up a slope inclined at 20° to the horizontal. The string is inclined at 30° to the slope. There is a frictional resistance of 40 N. The object accelerates up the slope at 0.8 m s$^{-2}$.
Find

**a** the tension in the string

**b** the normal reaction between the object and the slope.

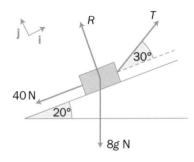

**a** Resolve parallel to the slope and use $F = ma$:

$$T\cos 30° - 40 - 8g\sin 20° = 8 \times 0.8$$

$$\Rightarrow \qquad\qquad T = 84.5 \text{ N}$$

**b** Resolving perpendicular to the slope there is no acceleration.

$$\Rightarrow \quad R + T\sin 30° - 8g\cos 20° = 0$$

Substitute for $T$ in this equation: $\quad R = 31.4$ N

In some situations you will need to use the equations relating to motion with constant acceleration.

Refer to 2.4 for a reminder of these equations.

EXAMPLE 3

A block of mass 5 kg moves on a rough horizontal plane with coefficient of friction 0.2 under the action of a horizontal force of 30 N. If the block starts from rest, find the distance it travels in the first 3 seconds of motion.

Let the block have acceleration $a$.

Use the model of friction:
$$F = 0.2R \qquad\qquad [1]$$

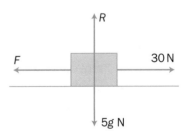

Resolve and apply Newton's laws:

Vertically $\qquad R - 5g = 0 \qquad [2]$
Horizontally $\quad 30 - F = 5a \qquad [3]$

Use equations [1] and [2] to find $F$:
$$F = 0.2 \times 5g = 9.8 \text{ N}$$

Substitute in equation [3]:
$$20.2 = 5a \quad \Rightarrow \quad a = 4.04 \text{ m s}^{-2}$$

Use $s = ut + \frac{1}{2}at^2$, where $u = 0$, $t = 3$ s and $a = 4.04$ m s$^{-2}$:

$$s = \frac{1}{2} \times 4.04 \times 9 = 18.2$$

So the block travels 18.2 m in the first 3 seconds of motion.

EXAMPLE 4

A particle of mass 6 kg is initially moving with a speed of $8 \text{ m s}^{-1}$ on a rough horizontal surface with a coefficient of friction 0.25. Find the distance it moves across the rough surface before coming to rest.

Let the particle have acceleration $a$.
Use the model of friction:

$$F = 0.25R \qquad [1]$$

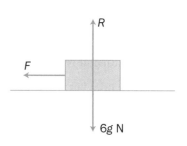

Resolve and apply Newton's laws:

Vertically $\qquad R - 6g = 0 \qquad [2]$
Horizontally $\qquad -F = 6a \qquad [3]$

Use equations [1] and [2] to find $F$:

$$F = 0.25 \times 6g = 14.7 \text{ N}$$

Substitute in equation [3]:

$$-14.7 = 6a \quad \Rightarrow \quad a = -2.45 \text{ m s}^{-2}$$

Use $v^2 = u^2 + 2as$, where $u = 8 \text{ m s}^{-1}$, $v = 0$ and $a = -2.45 \text{ m s}^{-2}$:

$$0 = 8^2 - 2 \times 2.45 \times s \quad \Rightarrow \quad s = 13.1$$

So, the block travels 13.1 m before coming to rest.

## Exercise 5.2

1   An object of mass 20 kg is moving vertically on the end of a cable. The only forces acting on the object are its weight and the tension in the cable.

   a   Find the acceleration of the object when the tension is

   i   250 N          ii   150 N.

   b   Find the tension in the cable when the object is

   i    moving upwards at a constant speed of $5 \text{ m s}^{-1}$
   ii   moving downwards at a constant speed of $4 \text{ m s}^{-1}$
   iii  accelerating upwards at $2 \text{ m s}^{-2}$
   iv   moving upwards and slowing uniformly from a speed of $6 \text{ m s}^{-1}$ to $2 \text{ m s}^{-1}$ in 6 seconds
   v    moving downwards and slowing uniformly from a speed of $6 \text{ m s}^{-1}$ to rest in 8 metres.

2   A car of mass 700 kg is acted upon by a driving force of 2200 N and a constant resistance of 800 N. The car starts from rest and travels along a horizontal road. After 6 seconds the driver depresses the clutch, stopping the driving force, and the car coasts to rest.

   a   What was the greatest speed achieved by the car?

   b   How far did the car travel altogether?

3 A block of mass 3 kg is initially moving at a speed of 10 m s$^{-1}$ on a rough horizontal surface with coefficient of friction 0.35.

   a Find the distance it travels before coming to rest.

   b By repeating the question for an object of mass $m$, show that the distance travelled by the object is independent of its mass.

4 A particle is moving at an initial speed of 6 m s$^{-1}$ on a rough horizontal surface and is brought to rest in a distance of 20 m. Find the coefficient of friction involved.

5 A block of mass 6 kg moves on a rough horizontal surface, coefficient of friction 0.25, under the action of a horizontal force. It accelerates from rest to a speed of 4 m s$^{-1}$ in a distance of 12 m, continues for a time at this speed, then decelerates to rest in a distance of 2 m. Find the magnitude and direction of the horizontal force required during each stage of the journey.

6 A moving particle encounters two rough areas, each 10 m wide. The coefficients of friction for the two areas are 0.2 and 0.4 respectively. Find the minimum initial speed of the particle if it makes it across the two areas.

7 A particle of mass 5 kg is being towed at a constant speed of 6 m s$^{-1}$ on a rough horizontal plane with coefficient of friction 0.2.
At a certain point the towing force is reversed in direction.

   a Find the distance the particle will travel before coming to rest.

   b If the force continues to act, what will happen after the particle has come to rest?

8 A block moves on a rough slope of length 10 m inclined at 30° to the horizontal. The coefficient of friction between the block and the slope is 0.4. The block starts from rest at the top of the slope.

   a Find the speed with which the block reaches the bottom of the slope.

   b The block is then projected back up the slope with initial speed $v$ m s$^{-1}$. It just reaches the top of the slope. Find $v$.

### Newton's third law of motion
For every action there is an equal and opposite reaction.

In other words, if an object *A* exerts a force on a second object *B* (by direct contact or at a distance by magnetic attraction, gravitation etc.), then *B* will exert a force on *A*. The two forces will be of equal magnitude and in opposite directions.

e.g. If two people of equal mass stand together on an ice rink and one pushes the other, they **both** start to move at equal speeds but in opposite directions.

If *A* and *B* are parts of the same system, the force of *A* on *B* and the force of *B* on *A* cancel out. They are forces internal to the system and do not affect the acceleration of the system. They are only taken into account if you want to calculate the acceleration of object *A* (or *B*) alone.

EXAMPLE 1

A man of mass 90 kg stands in a lift of mass 300 kg.
The cable of the lift has a tension of 4056 N.
Find the reaction between the man and the floor of the lift.

If you treat the man and the lift as one system, the reaction forces between the man and the floor are internal and can be ignored.

4056 N

In this example $g = 9.8 \text{ m s}^{-2}$.

Let the acceleration of the system be $a$ m s$^{-2}$.
Resolve upwards and use $F = ma$:
$$4056 - 390g = 390a$$
$$a = 0.6$$

390$g$ N

The system is accelerating upwards at 0.6 m s$^{-2}$.

To find the reaction between the man and the lift, consider the forces acting on the man.
Resolve upwards and use $F = ma$:
$$R - 90g = 90 \times 0.6$$
$$R = 936 \text{ N}$$

$R$

90$g$ N

By Newton's third law, the reaction of the floor on the man is equal and opposite to the reaction of the man on the floor.

Alternatively, you could consider the forces acting on the lift.
Resolve upwards and use $F = ma$:
$$4056 - 300g - R = 300 \times 0.6$$
$$R = 936 \text{ N}$$

4056 N

300$g$ N     $R$

EXAMPLE 2

An engine of mass 10 tonnes is pulling a truck of mass 3 tonnes. The resistance on the engine and the truck are 4000 N and 1500 N respectively. The driving force of the engine is 14 000 N. Find the acceleration of the system and the tension in the coupling between the engine and the truck.

There are vertical forces acting on the system. However, these have no effect on the horizontal motion so you can omit them from the diagram.

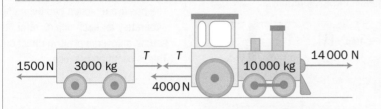

When finding the acceleration, $a$, you treat the engine and truck as one system. The tension in the coupling is internal to the system and so does not appear in the equations.

Consider the horizontal forces and use $F = ma$:

$$14\,000 - 4000 - 1500 = (10\,000 + 3000)a$$
$$\Rightarrow \qquad a = 0.654 \text{ m s}^{-2}$$

To find the tension in the coupling, you consider just the forces acting on the truck.

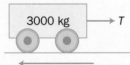

Consider the horizontal forces and use $F = ma$:

$$T - 1500 = 3000 \times 0.654$$
$$\Rightarrow \qquad T = 3460 \text{ N} \quad \text{(to 3 s.f.)}$$

Alternatively, you could consider just the forces acting on the engine:

$$14\,000 - 4000 - T = 10\,000 \times 0.654$$
$$\Rightarrow \qquad T = 3460 \text{ N}$$

Example 2 involved two objects connected together. You could call them connected particles. This term is also used to describe objects connected by strings passing over pulleys or other supports.

For problems involving pulleys, the usual modelling assumptions are that
○ the objects are particles, so you can ignore air resistance
○ the string is light, so its weight can be ignored
○ the string is inextensible, so the two ends of the string have the same speed and acceleration
○ the pulley is smooth, so the tension is the same throughout the string
○ the pulley is light, so no force is required to turn it.

M1

MI

EXAMPLE 3

Particles of mass 3 kg and 5 kg are attached to the ends of a light, inextensible string passing over a smooth pulley. The system is released from rest. Find the acceleration of the system and the tension in the string.

Write down the equation of motion for each mass separately:

For the 5 kg mass $\quad 5g - T = 5a \quad$ [1]
For the 3 kg mass $\quad T - 3g = 3a \quad$ [2]

Solve these simultaneous equations.

Add [1] and [2]:

$$2g = 8a \quad \Rightarrow \quad a = \frac{1}{4}g = 2.45 \text{ ms}^{-2}$$

Substitute this value for $a$ in equation [2]:

$$T - 3g = \frac{3}{4}g \quad \Rightarrow \quad T = 3\frac{3}{4}g = 36.75 \text{ N}$$

The positive direction has been chosen separately for each object, rather than setting an overall positive direction. This is common practice.

EXAMPLE 4

A block of mass 4 kg rests on a rough horizontal table, with coefficient of friction 0.5. It is attached by a light, inextensible string to a particle of mass 9 kg. The string passes over a smooth pulley at the edge of the table and the 9 kg mass hangs freely. Find the acceleration of the system, the tension in the string and the resultant force acting on the pulley.

Resolve vertically for the 4 kg mass:

$$R - 4g = 0 \quad \Rightarrow \quad R = 4g \text{ N}$$

The system will move, so the friction force will be at its maximum.

$$F = 0.5R = 2g \text{ N}$$

Write down the equations of motion for the two masses:

For the 9 kg mass: $\quad 9g - T = 9a \quad$ [1]
For the 4 kg mass: $\quad T - 2g = 4a \quad$ [2]

Add [1] and [2]: $\quad 7g = 13a \Rightarrow a = 5.28 \text{ m s}^{-2}$

Substitute this into equation [2]:

$$T - 2g = 21.1 \Rightarrow T = 40.7 \text{ N}$$

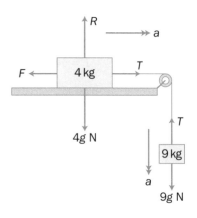

Drawing diagrams will help you to deal with more complicated situations.

To find the force on the pulley, you need to realise that each section of the string exerts a force on the pulley, as shown.

Find the resultant of these forces:

$$\text{Resultant force} = \sqrt{40.7^2 + 40.7^2} = 57.6 \text{ N}$$

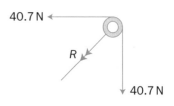

This resultant force acts at 45° to the horizontal.

EXAMPLE 5

The diagram shows a block $A$, of mass 2 kg, 3 m from the top of a smooth slope inclined at 30° to the horizontal. It is connected to a block $B$, of mass 3 kg, by a light inextensible string passing over a smooth pulley at the top of the slope. Block $B$ hangs freely a distance of 1 m above a horizontal plane. The system is released from rest.

Find

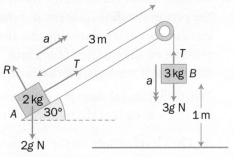

**a** the speed with which $B$ hits the plane

**b** how close $A$ gets to the top of the slope.

**a** Write down the equation of motion for each mass:

For the 3 kg mass $\qquad\qquad\qquad\qquad 3g - T = 3a \quad [1]$

For the 2 kg mass $\qquad\qquad\quad T - 2g \sin 30° = 2a \quad [2]$

Add [1] and [2]: $3g - 2g \sin 30° = 5a \Rightarrow a = 0.4g = 3.92 \text{ m s}^{-2}$

Use $v^2 = u^2 + 2as$ for $B$, with $u = 0 \text{ m s}^{-1}$, $s = 1$ m and $a = 3.92 \text{ m s}^{-2}$:

$\qquad v^2 = 7.84 \quad \Rightarrow \quad v = 2.8 \text{ m s}^{-1}$

**b** When $B$ hits the plane, the string goes slack.

Draw a new diagram to show the forces on $A$ now:

Let the acceleration of $A$ be $a_1$.

Consider the equation of motion for $A$:

$\qquad -2g \sin 30° = 2a_1$

$\Rightarrow \qquad\qquad a_1 = -0.5g = -4.9 \text{ m s}^{-2}$

Use $v^2 = u^2 + 2as$ for $A$, with $u = 2.8 \text{ m s}^{-1}$, $v = 0 \text{ m s}^{-1}$ and $a = -4.9 \text{ m s}^{-2}$:

$\qquad 0 = 7.84 - 9.8s \quad \Rightarrow \quad s = 0.8$ m

So $A$ travels 1 m before the string goes slack and another 0.8 m before coming to rest. Hence $A$ comes to rest 1.2 m from the top of the slope.

MI

## Exercise 5.3

1 An object of mass 50 kg is placed on the floor of a lift. Find the reaction between the object and the floor when the lift is

Use $g = 9.8 \text{ m s}^{-2}$

**a** accelerating upwards at $1.2 \text{ m s}^{-2}$

**b** moving upwards at a constant speed of $3.5 \text{ m s}^{-1}$

**c** moving upwards but slowing uniformly from $5 \text{ m s}^{-1}$ to $2 \text{ m s}^{-1}$ in 4 seconds

**d** accelerating downwards at $2 \text{ m s}^{-2}$.

2 Bathroom scales measure the reaction between the scales and the person standing on them, but the dial is calibrated to show the mass of the person assuming that the scales are placed in a horizontal position on the surface of the Earth. This means that, if the reaction is $R$ newtons, the dial is calibrated to show $(R \div 9.8)$ kg.

What will the dial show if a person of mass 80 kg stands on the scales in each of the following situations?

**a** On a level surface on the Moon, where the acceleration due to gravity is 1.6 m s$^{-2}$.

**b** On the horizontal floor of a lift moving upwards at a constant 3 m s$^{-1}$.

**c** On the horizontal floor of a lift accelerating upwards at 1.5 m s$^{-2}$.

**d** On the horizontal floor of a lift accelerating downwards at 0.8 m s$^{-2}$.

**e** On a surface sloping at 25° to the horizontal.

3 An object of mass 20 kg hangs from a spring balance in a lift. Its apparent mass is 24 kg. What is the acceleration of the lift?

4 An object hangs on a spring balance in a lift. The lift first accelerates upwards then accelerates downwards, both times with magnitude $a$ m s$^{-2}$. If the reading on the balance for the first phase was twice that for the second phase, find the value of $a$.

5 A concrete block of mass 80 kg rests on a pallet of mass 20 kg, which is attached to the cable of a crane, as shown. The crane lifts the pallet with an acceleration of 0.5 m s$^{-2}$. Find

**a** the tension in the cable

**b** the reaction between the block and the pallet.

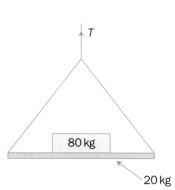

6 Two objects, of mass 3 kg and 4 kg, are connected by a light inextensible string, and both can be raised and lowered on the end of a second string, as shown. Find the tension in each of the strings when the system is

**a** at rest

**b** moving upwards at a constant speed of 2 m s$^{-1}$

**c** moving upwards with acceleration 3 m s$^{-2}$.

M1

7 A block of mass 5 kg is suspended by means of two identical light strings from a rod of mass 3 kg, with the strings making angles of 30° with the horizontal. The rod is suspended by another light string, as shown.

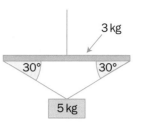

a Find the tension in each of the three strings if the system is accelerating upwards at 1.5 m s$^{-2}$.

b Each of the three strings has a breaking strain of 120 N. What is the maximum possible upward acceleration of the system, and which string will break if this maximum is exceeded?

8 A car of mass 800 kg is towing a caravan of mass 300 kg along a horizontal road. The constant resistances on the car and caravan are 700 N and 1200 N respectively.

a When the car exerts a driving force of 3000 N, find the acceleration of the system and the tension in the coupling.

b Find the force in the coupling when the car is towing the caravan at a constant speed of 50 km h$^{-1}$.

c Find the force in the coupling when the car exerts a braking force of 2000 N.

9 A block A, of mass 3 kg, is connected by means of a light string to a block B, of mass 4 kg. The blocks are placed on a rough horizontal surface with the string just taut. The coefficient of friction is 0.4 between block A and the surface, and 0.6 between block B and the surface. A horizontal force is applied to block A in the direction BA, causing the system to accelerate at 1.5 m s$^{-2}$.

a Find the magnitude of the applied force.

b Find the tension in the string.

10 Two particles, of mass 5 kg and 7 kg, are connected by a light, inextensible string passing over a smooth pulley. The particles are released with the two sections of the string taut and vertical. Find, in terms of g,

a the acceleration of the system

b the tension in the string

c the force on the pulley.

11 Two particles, of mass 2 kg and 3 kg, are connected by a light, inextensible string passing over a smooth pulley. The system is released from rest with the two sections of the string taut and vertical, and the 3 kg particle a distance of 4 m above the ground. Find the acceleration of the system and the speed with which the 3 kg particle hits the ground. Give your answers in terms of g.

M1

119

12 Two particles of mass $m$ and $2m$ are connected by a light, inextensible string passing over a smooth pulley. The particles are released with the two sections of the string taut and vertical. Find the acceleration of the system and the tension in the string in terms of $m$ and $g$.

13 A block of mass 3 kg rests on a smooth table. It is connected by a light, inextensible string passing over a smooth pulley at the edge of the table to a 2 kg particle hanging freely. Find the acceleration of the system and the tension in the string.

14 A block of mass 4 kg rests on a rough table. It is connected by a light, inextensible string passing over a smooth pulley at the edge of the table to a 5 kg particle hanging freely. There is a frictional force of 20 N acting on the block. Find the acceleration of the system and the tension in the string.

15 A block of mass 2 kg rests on a smooth table. It is connected by a light, inextensible string passing over a smooth pulley at the edge of the table to a 3 kg particle hanging freely. The block starts from rest a distance of 1.5 m from the pulley. Find the acceleration of the system and the time taken for the block to reach the pulley.

16 The diagram shows a block of mass 4 kg resting on a smooth plane inclined at 20° to the horizontal. It is connected by a light, inextensible string passing over a smooth pulley at the top of the slope to a particle of mass 3 kg hanging freely.

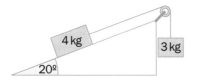

Find the acceleration of the system and the tension in the string.

17 Particles $A$ and $B$, of mass 2 kg and 5 kg respectively, are connected by a light, inextensible string passing over a smooth pulley. Initially the system is at rest with $A$ on the ground and $B$ 3 m above the ground. The string is taut. The system is released.

a Find the acceleration of the system.

b Find the speed with which the system is moving when $B$ hits the ground.

c How much further will $A$ rise before coming instantaneously to rest?

18 The diagram shows an object *A* of mass 5 kg connected by a light, inextensible string passing over a smooth pulley to a box *B* of mass 4 kg. There is an object *C* of mass 2 kg resting on the horizontal floor of the box. Find

a the acceleration of the system

b the reaction between *C* and the floor of the box.

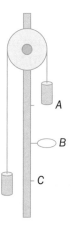

19 The diagram shows a version of what is known as Attwood's machine, which was used as a means of estimating the value of *g*. Two objects, both of mass *M*, are connected by a light string passing over a smooth pulley. The system starts from rest with one of the masses at *A*, as shown, and with a small rider of mass *m* attached to it. The system moves through a distance *h*, at which point the mass passes through a ring *B* which removes the rider. The system continues to move at uniform speed and the mass is timed in its descent from *B* to *C*, a distance *h*. If the system takes a time *t* to move from *B* to *C*, show that

$$g = \frac{h(2M + m)}{2mt^2}$$

20

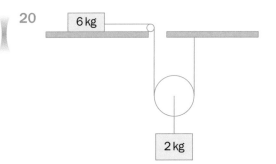

The diagram shows a block of mass 6 kg on a smooth horizontal table. A light inextensible string is attached to the block. It passes over a smooth pulley at the edge of the table, then under a second light smooth pulley which is free to move, and is finally fixed at the same level as the table. The moveable pulley carries a load of 2 kg.

a The system is released from rest and the 6 kg block has acceleration *a* m s$^{-2}$. State the acceleration of the 2 kg block.

b Write down equations of motion for the two blocks. Hence find the value of *a* and the tension in the string.

M1

1 A car of mass 1200 kg moves under the action of a driving force $P$ N and a resistance of 600 N. The car starts from rest and accelerates to a speed of 20 m s$^{-1}$ in a time of 8 s, after which it moves at constant speed. Calculate the value of $P$

   a  while the car is accelerating

   b  while the car travels at constant speed.

2 Find the acceleration of a body of mass 6 kg acted upon by forces of $(4\mathbf{i} + 7\mathbf{j})$ N, $(2\mathbf{i} - 3\mathbf{j})$ N and $(6\mathbf{i} + 8\mathbf{j})$ N.

3 A block of mass 2 kg is towed across a rough horizontal floor by means of a rope. The coefficient of friction between the block and the floor is 0.3, and the acceleration of the block is 0.5 m s$^{-2}$. Find the tension in the rope

   a  if the rope is horizontal

   b  if the rope is inclined at 20° to the horizontal.

4 A block of mass 10 kg rests on a rough horizontal surface. The coefficient of friction between the block and the surface is 0.4. A horizontal force, $P$ N, is applied to the block so that it accelerates from rest at 0.8 m s$^{-2}$.

   a  Find the value of $P$.

   b  When the block reaches a speed of 8 m s$^{-1}$, the force $P$ is removed and the block slows to rest. Find the total distance travelled by the block.

5 A particle of mass 4 kg is acted upon by three forces, $(3\mathbf{i} + 14\mathbf{j})$ N, $(7\mathbf{i} - 9\mathbf{j})$ N and $\mathbf{F}$. Find the force $\mathbf{F}$

   a  when the particle moves with constant velocity $(2\mathbf{i} + 3\mathbf{j})$ m s$^{-1}$

   b  when the particle moves with acceleration $(4\mathbf{i} + 6\mathbf{j})$ m s$^{-2}$.

6 A car of mass 1000 kg tows a trailer of mass 200 kg along a horizontal road. The resistances to motion are 400 N acting on the car and 80 N acting on the trailer.

   a  The car exerts a driving force of 1920 N.
   Calculate   i   the acceleration of the system
   ii  the tension in the coupling between the car and the trailer.

   b  Find the force in the coupling if
   i   the car and trailer are travelling at a constant speed
   ii  the driving force is removed and the car and trailer coast to rest.

7 Two particles, of mass 3 kg and 4 kg, are connected by a light inextensible string which passes over a smooth pulley. The system is released from rest with the particles hanging vertically. Find, in terms of $g$,

   a the acceleration of the system

   b the tension in the string.

8 A block of mass $3m$ kg rests on a rough horizontal table. The coefficient of friction between the block and the table is 0.2. A light inextensible string attached to the block passes over a smooth pulley at the edge of the table. A second block, of mass $m$ kg, hangs freely from the other end of the string. The system is released from rest. Find, in terms of $m$ and $g$,

   a the acceleration of the system

   b the tension in the string.

9 A particle $P$ of mass 1.5 kg is moving under the action of a constant force $(3\mathbf{i} - 7.5\mathbf{j})$ N. Initially $P$ has velocity $(2\mathbf{i} + 3\mathbf{j})$ m s$^{-1}$. Find

   a the magnitude of the acceleration of $P$

   b the velocity of $P$, in terms of $\mathbf{i}$ and $\mathbf{j}$, when $P$ has been moving for 4 seconds.                                  [(c) Edexcel Limited 2002]

10

3 kg          $m$ kg

Two particles have mass 3 kg and $m$ kg, where $m < 3$. They are attached to the ends of a light inextensible string. The string passes over a smooth fixed pulley. The particles are held in position with the string taut and the hanging parts of the string vertical, as shown in the diagram. The particles are then released from rest.

The initial acceleration of each particle has magnitude $\frac{3}{7}g$. Find

   a the tension in the string immediately after the particles are released

   b the value of $m$.                                          [(c) Edexcel Limited 2004]

**11**

This diagram shows a lorry of mass 1600 kg towing a car of mass 900 kg along a straight horizontal road. The two vehicles are joined by a light towbar which is at an angle of 15° to the road. The lorry and the car experience constant resistances to motion of magnitude 600 N and 300 N respectively. The lorry's engine produces a constant horizontal force on the lorry of magnitude 1500 N. Find

**a** the acceleration of the lorry and the car

**b** the tension in the towbar.

When the speed of the vehicles is 6 m s$^{-1}$, the towbar breaks. Assuming that the resistance to the motion of the car remains of constant magnitude 300 N,

**c** find the distance moved by the car from the moment the towbar breaks to the moment when the car comes to rest.

**d** State whether, when the towbar breaks, the normal reaction of the road on the car is increased, decreased or remains constant. Give a reason for your answer.

[(c) Edexcel Limited 2005]

**12** A fixed wedge has two plane faces, each inclined at 30° to the horizontal. Two particles, $A$ and $B$, of mass $3m$ and $m$ respectively, are attached to the ends of a light inextensible string. Each particle moves on one of the plane faces of the wedge. The string passes over a small smooth light pulley fixed at the top of the wedge. The face on which $A$ moves is smooth. The face on which $B$ moves is rough. The coefficient of friction between $B$ and this face is $\mu$. Particle $A$ is held at rest with the string taut. The string lies in the same vertical plane as lines of greatest slope on each plane face of the wedge, as shown in the diagram.

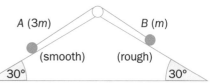

The particles are released from rest and start to move. Particle $A$ moves downwards and $B$ moves upwards. The accelerations of $A$ and $B$ each have magnitude $\frac{1}{10}g$.

**a** By considering the motion of $A$, find, in terms of $m$ and $g$, the tension in the string.

**b** By considering the motion of $B$, find the value of $\mu$.

**c** Find the resultant force exerted by the string on the pulley, giving its magnitude and direction.

[(c) Edexcel Limited 2006]

13

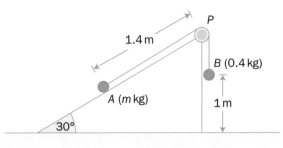

The diagram shows two particles A and B, of mass m kg and 0.4 kg respectively, connected by a light inextensible string. Initially A is held at rest on a fixed smooth plane inclined at 30° to the horizontal. The string passes over a small light smooth pulley P fixed at the top of the plane. The section of the string from A to P is parallel to a line of greatest slope of the plane. The particle B hangs freely below P. The system is released from rest with the string taut and B descends with acceleration $\frac{1}{5}g$.

**a** Write down an equation of motion for $B$.

**b** Find the tension in the string.

**c** Prove that $m = \frac{16}{35}$.

**d** State where in the calculations you have used the information that $P$ is a light smooth pulley.

On release, $B$ is at a height of one metre above the ground and $AP = 1.4$ m. The particle $B$ strikes the ground and does not rebound.

**e** Calculate the speed of $B$ as it reaches the ground.

**f** Show that $A$ comes to rest as it reaches $P$.

[(c) Edexcel Limited 2003]

M1

## Summary

Refer to

- Newton's three laws of motion are

  - **First law**

    Every object remains at rest or moves with constant
    velocity unless an external force is applied.

    5.1

  - **Second law**

    When an applied force causes an object to accelerate

    - the force and acceleration are in the same direction
    - the magnitude of the force is proportional to the
      magnitude of the acceleration and to the mass
      of the object.

      5.1, 5.2

      - Using standard units the second law gives the equation
        of motion $\mathbf{F} = m\mathbf{a}$ where $\mathbf{F}$ is the resultant force.

  - **Third law**

    For every action there is an equal and opposite reaction.

    5.3

- Forces internal to a system do not appear in the equation of
  motion of the whole system.
  To find them, you consider the equation of motion
  of one part of the system.

  5.3

- When particles are connected by a string passing over a pulley,
  you usually assume

  - the string is light and the pulley is smooth
  - the string is inextensible.

  You can write down the equations of motion of the particles
  separately, and solve to find the tension and the acceleration.

  5.3

### Links

The motion of an airboat relies on Newton's third law
– for every action there is an equal and opposite reaction.
The propeller at the back of the boat throws air molecules
backwards and the reaction force to this action pushes the
boat forwards. An airboat can be made more efficient by
altering the size of the propeller, the angle of its blade and
the r.p.m of the engine.

Rockets work using a similar principle. However, since there
is no air in space, they throw out the vaporised rocket fuel
backwards to create their forward thrust.

# Impulse and momentum

This chapter will show you how to
- define impulse and momentum
- use the relationship between impulse and momentum
- use the principle of conservation of momentum to solve
  problems involving collisions between particles.

## Before you start

### You should know how to:

1 Resolve a vector into components.

3.2 2 Manipulate vectors in component form.

2.4 3 Use the formulae relating to motion with
constant acceleration.

4 State and use Newton's laws of motion.

### Check in

1 A ship is travelling on a bearing of 050°
at 20 m s$^{-1}$. Taking east and north as the
**i**- and **j**-directions, express the velocity
of the ship in component form.

2 Find the values of $a$ and $b$ for which
$a(2\mathbf{i} + 3\mathbf{j}) + (\mathbf{i} - \mathbf{j}) = 6\mathbf{i} + b\mathbf{j}$.

3 A particle travelling at a speed of 8 m s$^{-1}$ has
an acceleration of –0.5 m s$^{-2}$. How far will it
go before coming to rest?

4 The particle in question 3 has a mass of
5 kg. Find the resultant force acting on
the particle as it slows down.

If you apply a force to an object, the effect it has depends on the mass of the object and for how long you exert the force.

You apply a constant force $F$ N for a time $t$ s in the direction of motion to a particle of mass $m$ kg.
The particle accelerates at $a$ m s$^{-2}$.
The velocity of the particle increases from $u$ m s$^{-1}$ to $v$ m s$^{-1}$.

You have $\qquad v = u + at$

$\Rightarrow \qquad a = \dfrac{v - u}{t}$

From Newton's second law you know that

$$F = ma$$

$\Rightarrow \qquad F = \dfrac{m(v - u)}{t}$

or $\qquad Ft = mv - mu$

This relationship defines two quantities.
The left-hand side, $Ft$, defines the impulse of the force.

The impulse of a force $\mathbf{F}$ applied for a time $t$ is defined as $\mathbf{F}t$.

Impulse = force × time

The SI unit of impulse is the newton second (N s).

Impulse is a vector quantity, because the force $\mathbf{F}$ is a vector.

The right-hand side, $mv - mu$, shows the change in value of the quantity 'mass × velocity'. This quantity is the momentum of the particle. Its units are the same as those of impulse.

This quantity is called **linear momentum** to distinguish it from angular momentum, but it is usual to just use the term momentum. Angular momentum is not in the M1 specification.

For a particle of mass $m$ moving with velocity $\mathbf{v}$,

Momentum = $m\mathbf{v}$

The SI unit of momentum is the newton second (N s).

Momentum is a vector quantity, because the velocity $\mathbf{v}$ is a vector.

Mass must be in kg and velocity must be in m s$^{-1}$.

In general,

> $\mathbf{F}t = m\mathbf{v} - m\mathbf{u}$
> Impulse = change of momentum

EXAMPLE 1

A spacecraft of mass 120 kg is travelling in a straight line at a speed of 4 m s$^{-1}$. Its rocket is fired for 5 seconds, exerting a force of 150 N. Find the new velocity of the spacecraft if the force was directed

**a** in the direction of motion

**b** in the direction opposite to the motion.

**a** The impulse $= 150 \times 5$
$= 750$ N s
The initial velocity $= 4$ m s$^{-1}$,
so the initial momentum $= 120 \times 4$
$= 480$ N s.
Let the final velocity be $v$.
The final momentum $= 120v$.

Impulse = change of momentum,
so $\quad 750 = 120v - 480$
$\Rightarrow \quad\quad v = 10.25$ m s$^{-1}$

**b** The situation is the same as in part **a** except that the impulse is now $-750$ N s.

Hence, $-750 = 120v - 480$
$\Rightarrow \quad\quad\quad v = -2.25$ m s$^{-1}$

That is, the spacecraft is now travelling at 2.25 m s$^{-1}$ in the opposite direction.

In reality firing the rocket would change the mass of the spacecraft because some fuel is burnt. The solution given here assumes that this change of mass is not significant.

M1

The concept of impulse is most useful when the period of time for which the force acts is very short.

e.g. When you hit a cricket ball with a bat, you cannot easily measure either the force (which is probably not constant) or the time for which it acts, but you can still calculate the change of momentum and hence the impulse.

If you can measure both the impulse and the time, you can find an average force, F, using impulse = Ft.

EXAMPLE 2

MI

A batsman hits a ball of mass 0.15 kg at a speed of 40 m s$^{-1}$. The ball moves in a straight line with constant speed until it strikes a net at right angles and is brought to rest in 0.5 s. Find the magnitude of the impulse and the average force exerted by the netting on the ball.

If you take left to right in the diagram as the positive direction, then the net exerts an impulse $-J$ N s on the ball, whose velocity changes from 40 m s$^{-1}$ to zero.

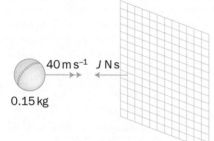

40 m s$^{-1}$   $J$ N s

0.15 kg

Initial momentum $= 0.15 \times 40$
$$= 6 \, \text{N s}$$
Final momentum $= 0 \, \text{N s}$

Impulse = change of momentum

so     $-J = 0 - 6$
$\Rightarrow$     $J = 6 \, \text{N s}$

So the netting exerts an impulse of magnitude 6 N s from right to left in the diagram.

To find the average force, $F$, use    Impulse = average force × time:

$$6 = F \times 0.5$$
$\Rightarrow$    $F = 12 \, \text{N}$

So the netting exerts an average force on the ball of 12 N from right to left in the diagram.

**EXAMPLE 3**

A particle of mass 2 kg, travelling with a velocity of $(3\mathbf{i} + 5\mathbf{j})$ m s$^{-1}$, is given an impulse of $(2\mathbf{i} - 4\mathbf{j})$ N s. Find its new velocity.

Let the new velocity be $\mathbf{v}$ m s$^{-1}$.

$$\text{Initial momentum} = 2(3\mathbf{i} + 5\mathbf{j})$$
$$= (6\mathbf{i} + 10\mathbf{j}) \text{ N s}$$
$$\text{Final momentum} = 2\mathbf{v} \text{ N s}$$

To find $\mathbf{v}$, use    Impulse = change of momentum:

$$2\mathbf{i} - 4\mathbf{j} = 2\mathbf{v} - (6\mathbf{i} + 10\mathbf{j})$$
$$\Rightarrow \qquad \mathbf{v} = 4\mathbf{i} + 3\mathbf{j}$$

The new velocity $= (4\mathbf{i} + 3\mathbf{j})$ m s$^{-1}$

**EXAMPLE 4**

Steve kicks a ball of mass 0.8 kg along the ground at a velocity of 5 m s$^{-1}$ towards Monica. She kicks it back towards him, but lofts it so that it leaves her foot with a speed of 8 m s$^{-1}$ and with an elevation of 40° to the horizontal. Find the magnitude and direction of the impulse of Monica's kick.

Take the unit vectors $\mathbf{i}$ and $\mathbf{j}$, as shown in the diagram.

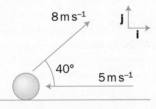

$$\text{Initial momentum} = 0.8 \times (-5\mathbf{i}) = -4\mathbf{i} \text{ N s}$$
$$\text{Final momentum} = 0.8 \times (8 \cos 40°\mathbf{i} + 8 \sin 40°\mathbf{j})$$
$$= (4.90\mathbf{i} + 4.11\mathbf{j}) \text{ N s}$$

Let the impulse be $\mathbf{J}$.

Use Impulse = change of momentum:

$$\mathbf{J} = (4.90\mathbf{i} + 4.11\mathbf{j}) - (-4\mathbf{i})$$
$$= (8.90\mathbf{i} + 4.11\mathbf{j}) \text{ N s}$$

This gives

$$\text{Magnitude of } \mathbf{J} = \sqrt{8.90^2 + 4.11^2} = 9.81 \text{ N s}$$

The direction of $\mathbf{J}$ is at an angle $\theta$ to the horizontal where

$$\tan \theta = \frac{4.11}{8.90} \qquad \Rightarrow \qquad \theta = 24.8°$$

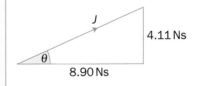

M1

## Exercise 6.1

1  An object of mass 4 kg is at rest. It receives an impulse of magnitude 28 N s. With what speed will it start to move?

2  An object of mass 7 kg is travelling in a straight line at a speed of 4 m s$^{-1}$. It is acted on by a constant force in the direction of the line, as a result of which its speed increases to 10 m s$^{-1}$.

   a  Find the impulse exerted on the object.

   b  Find the force involved if the event took 0.35 seconds.

3  A tennis player strikes a ball so that its path is exactly reversed. The ball approaches the racket at a speed of 35 m s$^{-1}$ and leaves at a speed of 45 m s$^{-1}$. The mass of the ball is 90 grams. Find the magnitude of the impulse exerted on the ball.

4  An engine of mass 20 tonnes is travelling at 54 km h$^{-1}$. Its brakes are applied for 3 s, after which it is travelling at 45 km h$^{-1}$. Find the change in momentum of the engine and hence the average braking force applied.

5  A particle of mass 3 kg travelling at 8 m s$^{-1}$ strikes a wall at right angles and receives an impulse of magnitude 39 N s. With what speed does it rebound from the wall?

6  An object of mass 2 kg has a velocity of $(8\mathbf{i} - 3\mathbf{j})$ m s$^{-1}$. It receives an impulse of $\mathbf{J}$ N s, which alters its velocity to $(2\mathbf{i} + 5\mathbf{j})$ m s$^{-1}$. Find $\mathbf{J}$.

7  A particle of mass 5 kg is travelling with a velocity of $(4\mathbf{i} + \mathbf{j})$ m s$^{-1}$ when it is subjected to an impulse of $(2\mathbf{i} - 7\mathbf{j})$ N s. Find the new velocity of the particle.

8  A particle of mass 3 kg, travelling with velocity $(2\mathbf{i} - 3\mathbf{j})$ m s$^{-1}$, is acted on by a constant force of $(-\mathbf{i} + 2\mathbf{j})$ N which changes its velocity to $0.5\mathbf{i}$ m s$^{-1}$. For how long does the force act?

9 A jet of water issues from a nozzle which has a square cross-section 2 cm by 2 cm. The water emerges at 10 m s$^{-1}$ and strikes a wall at right angles.

**a** Given that 1 m$^3$ of water has a mass of 1000 kg, calculate the mass of water striking the wall in one second.

**b** Making the modelling assumption that the water is brought to rest by striking the wall, calculate the change in momentum of 'one second's worth' of water.

**c** The wall exerts a force $F$ N on the water which, by Newton's third law, exerts an equal and opposite force on the wall. Use your answer from part **b** to find the value of $F$.

10 A rectangular block of mass 4 kg rests on a rough horizontal surface, with coefficient of friction 0.4. A jet of water, emerging from a circular nozzle of radius 6 cm with speed $v$ m s$^{-1}$, is sprayed directly on one of the vertical surfaces of the block. Assume that the water is brought to rest by striking the block.

**a** Find the value of $v$ for which the block will be on the point of moving.

**b** Find the initial acceleration of the block if the speed of the jet is twice that in part **a**.

11 A bullet of mass $m$, travelling east at a speed of $2u$, strikes an obstacle and ricochets. It continues to travel horizontally but on a bearing of 030° and its speed is halved.

**a** Taking east and north as the **i**- and **j** directions, write down the velocity of the bullet before and after the impact.

**b** Find, in terms of $m$ and $u$, the magnitude of the impulse sustained by the bullet.

**c** Find the direction of the impulse sustained by the bullet.

M1

You will now study the motion of systems comprising of two or more bodies, which are free to move separately.

A system consists of two bodies, $A$ and $B$, which have momentum $M_A$ and $M_B$ respectively.

The total momentum of the system $= (M_A + M_B)$ N s.

Body $A$ now exerts a force $F$ N on body $B$ for a time $t$ s. Body $B$ receives an impulse $J = Ft$ N s. By Newton's third law, $B$ exerts a force $-F$ N on $A$, also for time $t$ s. Body $A$ receives an impulse $-J$ N s.

The force might result from a collision, or be a magnetic or gravitational force. The bodies might be connected by a slack string which becomes taut and creates an impulsive force.

You know that
Impulse = change of momentum.
The new momentum of $A = (M_A - J)$ N s
The new momentum of $B = (M_B + J)$ N s

Even if the two forces are not constant, they are equal and opposite at all times, so the changes in momentum are equal and opposite.

The total momentum of the system $= (M_A - J) + (M_B + J)$
$$= (M_A + M_B) \text{ N s}$$

The total momentum has not been changed.

The forces described here are internal to the system. The total momentum of a system can only be changed by the application of an external force.

This is the principle of conservation of linear momentum.

**The principle of conservation of linear momentum**
The total momentum of a system in a particular direction remains constant unless an external force is applied in that direction.

Conservation of momentum is useful in solving problems involving collisions between objects.

EXAMPLE 1

A particle of mass 4 kg, travelling at a speed of 6 m s$^{-1}$, collides with a second particle, of mass 3 kg and travelling in the opposite direction at a speed of 2 m s$^{-1}$. After the collision, the first particle continues in the same direction but with its speed reduced to 1 m s$^{-1}$.
Find the velocity of the second particle after the collision.

Take left to right to be the positive direction.
Let the velocity of the second particle after the collision be $v$.
A sketch will help you to visualise the situation.

**Before collision**                **After collision**

Total momentum before collision $= 4 \times 6 + 3 \times (-2)$
$$= 18 \, \text{N s}$$
Total momentum after collision $= 4 \times 1 + 3v$
$$= (3v + 4) \, \text{N s}$$

There are no external forces, so momentum is conserved.
Therefore,             $3v + 4 = 18$
$$\Rightarrow \qquad v = 4\tfrac{2}{3} \, \text{m s}^{-1}$$

EXAMPLE 2

A body $A$ of mass 5 kg, travelling with a speed of 6 m s$^{-1}$, collides with a body $B$, of mass 3 kg, travelling in the same direction with a speed of 4 m s$^{-1}$. On colliding, the two bodies coalesce (merge into a single body).
Find the velocity of the combined body after the collision.

Let the velocity of the combined body after impact be $v$.
Sketch the situation:

**Before collision**                **After collision**

6 m s$^{-1}$       4 m s$^{-1}$              $v$
(5 kg)          (3 kg)                (8 kg)

Total momentum before collision $= 5 \times 6 + 3 \times 4 = 42 \, \text{N s}$
Momentum after collision $= 8v \, \text{N s}$

There are no external forces, so momentum is conserved.
Therefore,             $8v = 42$
$$\Rightarrow \qquad v = 5.25 \, \text{m s}^{-1}$$

A practical situation in which bodies 'coalesce' is when railway trucks collide and become coupled together.

EXAMPLE 3

A railway truck of mass 4 tonnes, travelling along a straight horizontal rail at a speed of 4 m s$^{-1}$, meets another truck, of mass 2 tonnes, travelling in the opposite direction at a speed of 5 m s$^{-1}$. They collide and become coupled together. Find their velocity after the collision.

Take left to right to be the positive direction.
Let the combined velocity after collision be $v$.

Total momentum before collision
$$= 4000 \times 4 + 2000 \times (-5)$$
$$= 6000 \text{ N s}$$

Total momentum after collision
$$= 6000v \text{ N s}$$

There are no external forces in the direction of motion, so momentum is conserved. Therefore,

$$6000v = 6000$$
$$\Rightarrow \quad v = 1 \text{ m s}^{-1}$$

**Before collision**

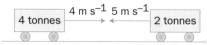

**After collision**

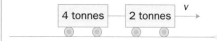

You can also use conservation of momentum in two-dimensional problems.

EXAMPLE 4

A body of mass 4 kg travelling with velocity $(3\mathbf{i} + 2\mathbf{j})$ m s$^{-1}$ collides and coalesces with a second body of mass 3 kg travelling with velocity $(\mathbf{i} - 3\mathbf{j})$ m s$^{-1}$. Find their common velocity after impact.

Let the common velocity after impact be $\mathbf{v}$.

Total momentum before collision $= 4(3\mathbf{i} + 2\mathbf{j}) + 3(\mathbf{i} - 3\mathbf{j})$
$$= (15\mathbf{i} - \mathbf{j}) \text{ N s}$$

Total momentum after collision $= 7\mathbf{v}$

By the principle of conservation of momentum,

$$7\mathbf{v} = 15\mathbf{i} - \mathbf{j}$$
$$\Rightarrow \quad \mathbf{v} = \left(2\tfrac{1}{7}\mathbf{i} - \tfrac{1}{7}\mathbf{j}\right) \text{ m s}^{-1}$$

Impulses between objects can occur in ways other than a collision.

Two particles, *A* and *B*, of mass 3 kg and 2 kg respectively, lie at rest on a smooth horizontal table. They are connected by a light inextensible string which is initially slack. *B* starts to move at a speed of 8 m s⁻¹ in the direction *AB*. Find the common velocity of the particles immediately after the string becomes taut, and the impulse received by each of the particles.

Let the common velocity be *v*.
Total momentum before = $2 \times 8 = 16$ N s
Total momentum after = $2v + 3v = 5v$ N s

There are only internal forces acting, so momentum is conserved. Therefore,

$$5v = 16 \quad \Rightarrow \quad v = 3.2 \text{ m s}^{-1}$$

The impulse received by each particle is given by its change of momentum.

For particle *A*  Initial momentum = 0
Final momentum = $3 \times 3.2 = 9.6$ N s
Impulse = $9.6 - 0 = 9.6$ N s

For particle *B*  Initial momentum = $2 \times 8 = 16$ N s
Final momentum = $2 \times 3.2 = 6.4$ N s
Impulse = $6.4 - 16 = -9.6$ N s

**Before string goes taut**

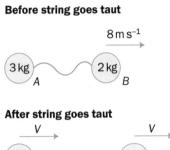

**After string goes taut**

The impulses on the particles are equal and opposite.

If the impulses result from a string becoming taut, you refer to the **impulsive tension** in the string.

When a gun is fired, the explosion exerts a forward force on the bullet and an equal backward force on the gun. If the gun is free to move it makes a sudden backward movement – the recoil. These explosive forces are internal to the system, so the total momentum of the gun and the bullet is unchanged.

The gun and bullet would usually be stationary before firing, in which case the total momentum is zero both before and after the shot is fired.

A bullet of mass 50 grams is fired horizontally from a gun of mass 1 kg, which is free to move. The bullet is fired with a velocity of 250 m s⁻¹. Find the speed with which the gun recoils.

The gun and the bullet are stationary to start with, so the total initial momentum is zero.

Total momentum after firing = $0.05 \times 250 + 1 \times (-v) = 12.5 - v$

The total momentum is conserved, so $12.5 - v = 0$

$$\Rightarrow \qquad v = 12.5 \text{ m s}^{-1}$$

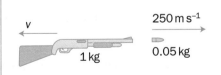

You can often make the modelling assumption that the impact or explosion occurs over such a short period of time that the velocity changes are effectively instantaneous. If there are external forces, such as friction, they will have an insignificant effect during the short time that the impulse acts.

EXAMPLE 7

A gun, of mass 800 kg, fires a shell of mass 4 kg horizontally at a speed of 400 m s$^{-1}$. The gun rests on a rough horizontal surface with coefficient of friction 0.6. The gun is stationary before the shot is fired. Find the distance that the gun will move as a result of firing the shell.

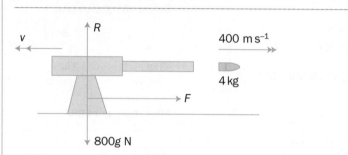

The gun and shell are stationary before firing, so the total initial momentum is zero.

Take left to right as the positive direction, and let the recoil speed be $v$ m s$^{-1}$.

$$\text{Total momentum after firing} = 4 \times 400 + 800 \times (-v)$$
$$= 1600 - 800v \text{ N s}$$

Assuming that the frictional force is negligible in comparison to the explosive forces, you can use the principle of conservation of momentum.

You have $\quad 1600 - 800v = 0$

$$\Rightarrow \qquad\qquad v = 2 \text{ m s}^{-1}$$

You now need to find the frictional force, $F$, which will bring the gun to rest after it has been fired.

Resolve vertically: $\quad R - 800g = 0 \quad \Rightarrow \quad R = 800g$ N

As the gun is moving, $F = \mu R = 0.6R \Rightarrow F = 480g$ N

Let the acceleration of the gun be $a$ m s$^{-2}$.

Consider the horizontal motion and use Newton's second law:
$$480g = 800a \quad \Rightarrow \quad a = 0.6g \text{ m s}^{-2}$$

After firing, the gun has an initial velocity of $-2$ m s$^{-1}$, a final velocity of zero and an acceleration of $0.6g$ m s$^{-2}$.

Use $v^2 = u^2 + 2as$:
$$0 = 4 + 1.2gs$$
$$\Rightarrow \quad s = -\frac{4}{1.2g} = -0.34 \text{ m}$$

So the gun moves 0.34 m to the left.

## Exercise 6.2

1  A bullet of mass 40 grams is fired horizontally with velocity $600 \text{ m s}^{-1}$ into a block of wood of mass 6 kg, which is resting on a smooth horizontal surface. The bullet becomes embedded in the block. Find the resulting common speed of the bullet and the block.

2  Two particles, $A$ and $B$, have masses of 2 kg and 3 kg respectively. They are travelling at speeds of $5 \text{ m s}^{-1}$ and $2 \text{ m s}^{-1}$ respectively. They collide and coalesce. Find their common speed after the collision if, before they collided, they were travelling

   a  in the same direction

   b  in opposite directions.

3  Arthur balances a box on top of a wall and throws a snowball of mass 0.3 kg at it. The snowball strikes the box at a speed of $10 \text{ m s}^{-1}$ and sticks to it. The common speed of the box and the snowball immediately after the impact is $4 \text{ m s}^{-1}$. Find the mass of the box.

4  A railway truck of mass $3m$, travelling at a speed of $2v$, collides with another truck, of mass $4m$, travelling with a speed of $v$. The trucks become coupled together. Find, in terms of $v$, the common speed of the trucks after impact if they were travelling

   a  in the same direction before impact

   b  in opposite directions before impact.

5  A particle $A$, of mass 10 kg, is moving at a speed of $5 \text{ m s}^{-1}$ when it collides with a particle $B$, of mass $m$ kg, travelling in the opposite direction at a speed of $2 \text{ m s}^{-1}$. After the collision, $A$ travels in the same direction as before but with its speed reduced to $3 \text{ m s}^{-1}$.

   a  If $m = 3$, find the velocity of $B$ after the collision.

   b  Show that the value of $m$ cannot be greater than 4.

6  A sledgehammer of mass 6 kg, travelling vertically downwards at a speed of $20 \text{ m s}^{-1}$, strikes the top of a post of mass 2 kg and does not rebound.

   a  Find the common speed of the hammer and post immediately after impact.

   b  If the post is driven 15 cm into the ground by the impact, find the average resistance of the ground to the motion of the post.

M1

7  A bullet of mass 50 grams is fired horizontally into a wooden block of mass 4 kg, which rests on a rough horizontal surface. The coefficient of friction between the block and the surface is 0.4. As a result of the collision the block, with the bullet embedded in it, moves a distance of 10 m along the surface before coming to rest. Find the speed at which the bullet struck the block.

8  Particles of mass 3 kg and 5 kg are connected by an elastic rope and are held apart on a smooth horizontal surface with the rope stretched. The particles are released from rest and a short time later the smaller mass has a speed of 6 m s$^{-1}$. Find the speed of the larger mass at this time.

9  A gun of mass 500 kg, which is free to move horizontally, fires a shell of mass 5 kg horizontally at a speed of 200 m s$^{-1}$. Find the speed of recoil of the gun.

10  A gun of mass 400 kg fires a shell of mass 8 kg horizontally at a speed of 300 m s$^{-1}$. Find the restraining force, assumed constant, which will be needed to bring the gun to rest in a distance of 2 m.

11  An object of mass 2 kg and velocity $(2\mathbf{i} - \mathbf{j})$ m s$^{-1}$ strikes and coalesces with a second object of mass 3 kg and velocity $(4\mathbf{i} + 6\mathbf{j})$ m s$^{-1}$. Find their common velocity after impact.

12  An object of mass 4 kg, travelling with velocity $(5\mathbf{i} + 2\mathbf{j})$ m s$^{-1}$, is struck by a second object of mass 6 kg and velocity $\mathbf{v}$, which sticks to it. Their common velocity after impact is $(2\mathbf{i} - 4\mathbf{j})$ m s$^{-1}$. Find $\mathbf{v}$.

13  An object of mass 3 kg has velocity $(3\mathbf{i} + 2\mathbf{j})$ m s$^{-1}$. It collides with another object, which has a mass of 2 kg and a velocity of $(\mathbf{i} - \mathbf{j})$ m s$^{-1}$. After the impact the first object has a velocity of $(2\mathbf{i} + \mathbf{j})$ m s$^{-1}$. Find the velocity of the second object.

14  A particle of mass 4 kg has a velocity of $(\mathbf{i} - \mathbf{j})$ m s$^{-1}$. It collides with a second particle, which has a mass of $m$ kg and a velocity of $(2\mathbf{i} - 3\mathbf{j})$ m s$^{-1}$. The first particle is brought to rest by the impact, and the second particle has a velocity of $(4\mathbf{i} + a\mathbf{j})$ m s$^{-1}$ after the impact. Find the values of $m$ and $a$.

15 Particles $A$ and $B$, of masses 4 kg and 2 kg respectively, lie at rest on a smooth horizontal surface, and are connected by a slack, light, inextensible string. Particle $A$ is given an impulse of 20 Ns so that it moves directly away from $B$. Find their common speed after the string becomes taut.

16 $n$ railway trucks, each of mass $m$ kg, stand at rest on a horizontal track. There are gaps of $d$ m between the trucks. The end truck is set in motion with speed $v$ m s$^{-1}$ towards the others. Each resulting collision results in the trucks involved in that collision becoming coupled together.
Find

   a the speed of the trucks when all of them have become coupled

   b the length of time between the first truck starting to move and the final truck becoming coupled.

(you may assume that
$1 + 2 + 3 + \ldots + n = \frac{1}{2}n(n + 1)$)

17 Arnold (mass 50 kg) is standing on a trolley (mass 20 kg) which is stationary. Bernice is standing behind the trolley. She throws Arnold a package of mass 5 kg, which he catches while it is travelling horizontally at a speed of 4 m s$^{-1}$, causing the trolley to start moving forwards. He immediately throws the package horizontally back towards Bernice at a speed of 5 m s$^{-1}$. Find the speed of the trolley at the end of this process.

18 Particles $A$, of mass 1 kg, and $B$, of mass 2 kg, are connected by a light inextensible string of length 2 m. Particle $A$ rests at the edge of a smooth horizontal table. Particle $B$ rests on the surface of the table so that $AB = 1$ m and $AB$ is perpendicular to the edge of the table. Particle $A$ is then nudged over the edge so that it falls from rest. With what speed will $B$ be jerked into motion?

19 Particles $A$, of mass 2 kg, and $B$, of mass 1 kg, are connected by a light inextensible string of length 1 m. $B$ is connected by a similar but longer string to a third particle, $C$, of mass 2 kg. The string $BC$ passes over a smooth pulley. The system is released from rest with $A$ and $B$ together on the ground and the string over the pulley taut and with both portions vertical. Find the speed with which particle $A$ leaves the ground and the height to which it rises. (Assume that particle $C$ never reaches the ground.)

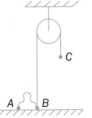

1 A batsman hits a ball so that its direction of travel is reversed. Before the blow the speed of the ball is 30 m s⁻¹ and afterwards it is 40 m s⁻¹. The mass of the ball is 120 grams. Find

   **a** the magnitude of the impulse exerted by the bat on the ball

   **b** the average force between bat and ball if they were in contact for 0.2 s.

2 A particle of mass 2 kg, travelling with velocity $(2\mathbf{i} - \mathbf{j})$ m s⁻¹, receives an impulse of $J$ N s. As a result its velocity is changed to $(5\mathbf{i} + 2\mathbf{j})$ m s⁻¹. Find $J$.

3

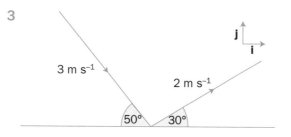

The diagram shows a snooker ball of mass 0.15 kg rebounding from a cushion.

   **a** Express the initial and final velocities of the ball in component form relative to the **i**- and **j**-directions shown.

   **b** Hence find the impulse exerted on the ball by the cushion.

4 A ball of mass 0.1 kg is dropped from rest 20 m above a horizontal surface. It strikes the surface and rebounds to a height of 14 m. Find the magnitude of the impulse exerted on the ball by the ground.

5 Two particles of equal mass are travelling along the same line at speeds of 5 m s⁻¹ and 7 m s⁻¹. They collide and coalesce. Find their common speed after impact if they were travelling

   **a** in opposite directions prior to impact

   **b** in the same direction prior to impact.

6  A bullet of mass 0.1 kg is fired horizontally at a block of wood of mass 2 kg, which is stationary and free to move. The bullet enters the block at a speed of 100 m s$^{-1}$. Find the subsequent speed of the block if

   a   the bullet passes through the block and emerges travelling at a speed of 40 m s$^{-1}$

   b   the bullet becomes embedded in the block.

7  A particle $A$, of mass 0.1 kg and travelling at a speed of 6 m s$^{-1}$, collides with a particle $B$, of mass 0.4 kg, which is at rest. After the impact the particles are travelling directly away from each other each with speed $V$ m s$^{-1}$. Find $V$.

8  Particles of mass $m$ and $2m$, both travelling with speed $u$, collide head on. After the collision one particle is travelling at twice the speed of the other. Find their speeds after impact in terms of $u$.

9  Two particles $A$ and $B$ have mass 0.12 kg and 0.08 kg respectively. They are initially at rest on a smooth horizontal table. Particle $A$ is then given an impulse in the direction $AB$ so that it moves with speed 3 m s$^{-1}$ directly towards $B$.

   a   Find the magnitude of this impulse, stating clearly the units in which your answer is given.

Immediately after the particles collide, the speed of $A$ is 1.2 m s$^{-1}$, its direction of motion being unchanged.

   b   Find the speed of $B$ immediately after the collision.

   c   Find the magnitude of the impulse exerted on $A$ in the collision.         [(c) Edexcel Limited 2003]

10  A railway truck $S$ of mass 2000 kg is travelling due east along a straight horizontal track with constant speed 12 m s$^{-1}$. The truck $S$ collides with a truck $T$ which is travelling due west along the same track as $S$ with constant speed 6 m s$^{-1}$. The magnitude of the impulse of $T$ on $S$ is 28 800 N s.

   a   Calculate the speed of $S$ immediately after the collision.

   b   State the direction of motion of $S$ immediately after the collision.

Given that, immediately after the collision, the speed of $T$ is 3.6 m s$^{-1}$, and that $T$ and $S$ are moving in opposite directions,

   c   calculate the mass of $T$.         [(c) Edexcel Limited 2003]

M1

11 A particle $P$ of mass 2 kg is moving with speed $u$ m s$^{-1}$ in a straight line on a smooth horizontal plane. The particle $P$ collides directly with a particle $Q$ of mass 4 kg which is at rest on the same horizontal plane. Immediately after the collision, $P$ and $Q$ are moving in opposite directions and the speed of $P$ is one-third the speed of $Q$.

   **a** Show that the speed of $P$ immediately after the collision is $\frac{1}{5}u$ m s$^{-1}$.

   After the collision $P$ continues to move in the same straight line and is brought to rest by a constant resistive force of magnitude 10 N. The distance between the point of collision and the point where $P$ comes to rest is 1.6 m.

   **b** Calculate the value of $u$.        [(c) Edexcel Limited 2004]

12 A tent peg is driven into soft ground by a blow from a hammer. The tent peg has mass 0.2 kg and the hammer has mass 3 kg. The hammer strikes the peg vertically. Immediately before the impact, the speed of the hammer is 16 m s$^{-1}$. It is assumed that, immediately after the impact, the hammer and the peg move together vertically downwards.

   **a** Find the common speed of the peg and the hammer immediately after the impact.

   Until the peg and hammer come to rest, the resistance exerted by the ground is assumed to be constant and of magnitude $R$ newtons. The hammer and peg are brought to rest 0.05 s after the impact.

   **b** Find, to three significant figures, the value of $R$.        [(c) Edexcel Limited 2004]

13 Two particles $A$ and $B$ have mass 0.4 kg and 0.3 kg respectively. They are moving in opposite directions on a smooth horizontal table and collide directly. Immediately before the collision, the speed of $A$ is 6 m s$^{-1}$ and the speed of $B$ is 2 m s$^{-1}$. As a result of the collision, the direction of motion of $B$ is reversed and its speed immediately after the collision is 3 m s$^{-1}$.
Find

   **a** the speed of $A$ immediately after the collision, stating clearly whether the direction of motion of $A$ is changed by the collision

   **b** the magnitude of the impulse exerted on $B$ in the collision, stating clearly the units in which your answer is given.        [(c) Edexcel Limited 2006]

**14 a** Two particles $A$ and $B$, of mass 3 kg and 2 kg respectively, are moving in the same direction on a smooth horizontal table when they collide directly. Immediately before the collision, the speed of $A$ is 4 m s$^{-1}$ and the speed of $B$ is 1.5 m s$^{-1}$. In the collision, the particles join to form a single particle $C$. Find the speed of $C$ immediately after the collision.

**b** Two particles $P$ and $Q$ have mass 3 kg and $m$ kg respectively. They are moving towards each other in opposite directions on a smooth horizontal table. Each particle has speed 4 m s$^{-1}$, when they collide directly. In this collision, the direction of motion of each particle is reversed. The speed of $P$ immediately after the collision is 2 m s$^{-1}$ and the speed of $Q$ is 1 m s$^{-1}$.
Find
**i** the value of $m$
**ii** the magnitude of the impulse exerted on $Q$ in the collision.

[(c) Edexcel Limited 2006]

M1

# 6

## Exit ⟹

### Summary

Refer to

- For a force **F** applied for a time $t$,

  Impulse = force × time = **F**$t$

  The SI unit of impulse is the newton second (N s)

  6.1

- For a particle of mass $m$ moving with velocity **v**,

  Momentum = $m$**v**

  The SI unit of momentum is the newton second (N s)

  6.1

- Impulse = change of momentum

  **F**$t = m$**v** − $m$**u**

  6.1

- The principle of conservation of linear momentum states that

  The total momentum of a system in a particular direction remains
  constant unless an external force is applied in that direction

  6.2

### Links

Conservation of momentum is crucial in many sports,
particularly snooker. The player hits the cue ball with the cue,
causing it to move. The cue ball then hits another ball which
then also moves. The reason for the movement of the balls is
due to conservation of momentum. The total momentum is
conserved before and after each collision. The player can,
however, control the momentum of each particular ball by
using different styles of cue action, forcing the balls to move in
a particular way.

MI

# 7

## Moments

This chapter will show you how to
- calculate the turning effect of a force about a given point
- solve problems involving parallel forces in equilibrium.

## Before you start

### You should know how to:

1 Resolve a force into components.

2 Use the trigonometry of right-angled triangles.

### Check in

1

Express each of these forces in component form relative to the **i**- and **j**-directions shown.

2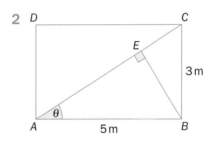

The diagram shows a rectangle $ABCD$, with $AB = 5$ m and $BC = 3$ m.
Calculate

  **a**  the angle $\theta$

  **b**  the length of $BE$.

When you apply a force to a particle, there is only one point at which the force can act.

With larger objects, applying the force at different points can have different effects.

The modelling assumption 'the object is a particle' implies that the object is so small that changing the point of application of the force has no significant effect.

e.g. When you close a door, it is easier if you push at the edge furthest from the hinges, and much harder if you push at a point close to the hinges. The strength of the turning effect depends not only on how hard you push, but also on where and in which direction you push – that is, it depends on the **line of action of the force**.

If the line of action passes through the hinge there will be no turning effect at all.

You can explore the situation experimentally.

### INVESTIGATION

Pivot a uniform rod at its centre, so that it can rotate in a vertical plane.

The rod in the diagram is 1 m long.

Hang a particle near one end. The diagram shows a 0.04 kg mass placed 0.48 m from the pivot. The downward force 0.04$g$ N gives an anticlockwise turning effect.

Place another particle, of mass $m$ kg, a distance $x$ m on the other side of the pivot. The downward force $F = mg$ gives a clockwise turning effect.

Adjust the value of $x$ so that the rod balances. The two turning effects are equal and opposite.

Here are some results for different values of $m$.

Try this experiment.

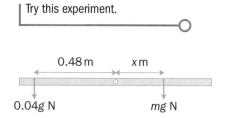

0.48 m      $x$ m

0.04$g$ N         $mg$ N

| Mass (kg) | Distance (m) | Force = $mg$ (N) | Force × Distance (N m) |
|---|---|---|---|
| $m$ | $x$ | $F$ | $Fx$ |
| 0.04 | 0.48 | 0.392 | 0.19 |
| 0.06 | 0.32 | 0.588 | 0.19 |
| 0.08 | 0.24 | 0.784 | 0.19 |
| 0.10 | 0.19 | 0.980 | 0.19 |
| 0.12 | 0.16 | 1.176 | 0.19 |
| 0.14 | 0.14 | 1.372 | 0.19 |
| 0.16 | 0.12 | 1.568 | 0.19 |
| 0.18 | 0.11 | 1.764 | 0.19 |

Each combination of $F$ and $x$ gives the same turning effect.

The product $Fx$ is the same every time.
This product is the **moment of the force**.
It gives the strength of the turning effect.

MI

## Moment of a force

A force **F** acts at a distance $d$ from a point $A$.
The turning effect of the force about
the point has magnitude $|F| \times d$.

The turning effect is called the
moment of the force or the torque.

As $|F|$ is in newtons and $d$ is in metres,
the unit for the moment of a force is the newton metre ($N\,m$).

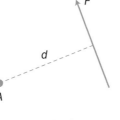

The turning effect is really about an axis through $A$ perpendicular to the plane containing **F**. However in two-dimensional problems it is usual to talk about the moment about a point.

You need to consider the direction (the sense) of the moment.

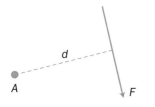

**a** Clockwise moment     **b** Anticlockwise moment

In these two diagrams the magnitude of the moment is the same, but they are in opposite senses.

You usually take anticlockwise as the positive rotational direction.
In the diagram, the moment in case **a** is $-F \times d$ and
in case **b** is $+F \times d$.

This convention is consistent with the positive and negative rotational directions that you may have met in your study of trigonometry.

In most situations you have two or more forces, each with its own turning effect. You can combine these to give an overall moment.

You can find the moment of a system of forces about a point by adding the moments of the individual forces.

Remember to allow for positive and negative moments.

**EXAMPLE 1**

Forces of 10 N, 15 N and 18 N act as shown on a rectangular lamina $ABCD$, with $AB = 6$ m and $BC = 4$ m.
Find the total moment of the forces about $A$.

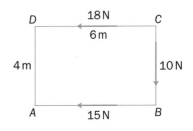

Remember that the term **lamina** refers to an idealised, infinitely thin plane object.

The 18 N force acts along a line 4 m from $A$ and turns in an anticlockwise sense.
Its moment is   $+18 \times 4 = +72$ N m

The 10 N force acts along a line 6 m from $A$ and turns in a clockwise sense.
Its moment is   $-10 \times 6 = -60$ N m

The 15 N force acts directly through A, so its moment is $0$ N m.

The total moment of the forces is   $72 - 60 + 0 = 12$ N m

EXAMPLE 2

The diagram shows a triangular lamina $ABC$, with $AB = AC = 4$ m. $D$ is the mid-point of $AC$. Forces of 6 N, 8 N, 12 N and $P$ act as shown. The total moment about $A$ is 10 Nm.

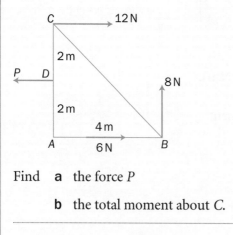

Find **a** the force $P$

**b** the total moment about $C$.

**a** The total moment about $A$ is

$$8 \times 4 + P \times 2 - 12 \times 4 = 2P - 16$$

Hence $2P - 16 = 10 \quad \Rightarrow \quad P = 13$ N

**b** The total moment about $C$ is

$$6 \times 4 + 8 \times 4 - P \times 2 = 56 - 2P = 30 \text{ N m}$$

## Exercise 7.1

**1** Find the moment of each of these forces about the point $A$, indicating whether the moment is positive or negative.

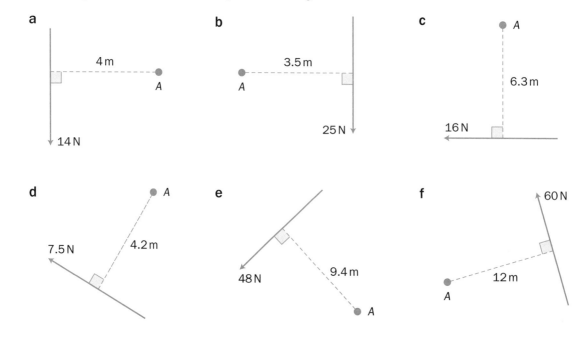

2 Find the total moment of the forces shown in each of these diagrams about　i the point *A*　ii the point *B*.

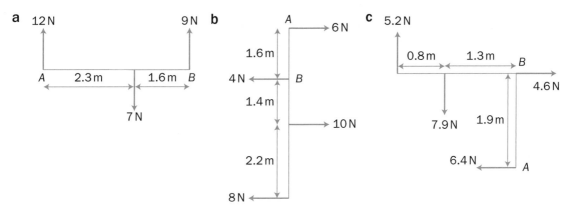

3 The rectangular lamina *ABCD* has *AB* = 4.8 m and *BC* = 3.6m. *M* is the mid-point of *AB* and *O* is the centre of the rectangle. Forces 2 N, 3 N, 4 N and 6 N act as shown in the diagram.
Find the sum of the moments of these forces about

a *A*　　　b *B*

c *M*　　　d *O*

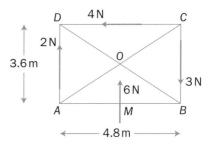

MI

4 The diagram shows a plan view of a revolving door. Four people are exerting forces of 40 N, 60 N, 80 N and 90 N as shown. Find the distance *x* if the total moment of the forces about *O* is

a 12 N m　　b –8 N m　　c 0 N m

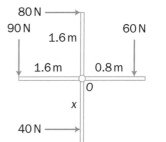

5 The diagram shows a framework *ABCD* made of light rods, each of length 2*a* m. Forces of 50 N, 80 N, 60 N and *P* act as shown. The total moment of the system of forces about *B* is –50*a* N m.

a Find the force *P*.

b Show that the total moment of the system is the same about all four points *A*, *B*, *C* and *D*.

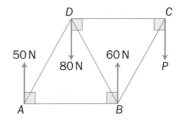

6 Forces of *P* N, *Q* N and 10 N act along the sides and diagonal of a rectangle, as shown. Given that the total moment of the system is the same about any point, find the values of *P* and *Q*.

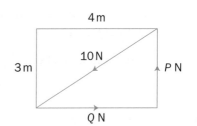

151

Suppose you have a rod $AB$ of length $a$. You apply a force of magnitude $F$ at $B$, at an angle $\theta$ to the rod. You want to find the moment of the force about $A$.

There are two possible methods.

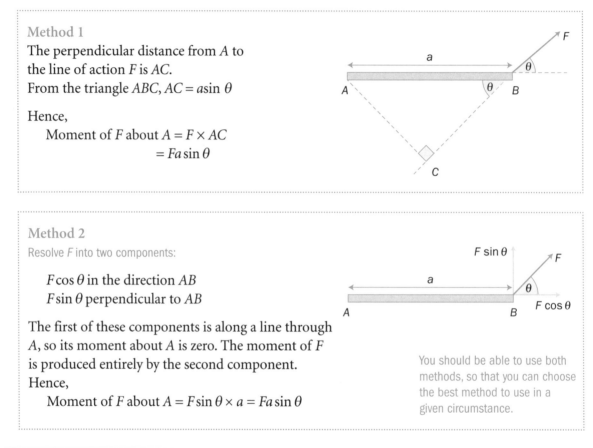

### Method 1

The perpendicular distance from $A$ to the line of action $F$ is $AC$.

From the triangle $ABC$, $AC = a\sin\theta$

Hence,

$\quad$ Moment of $F$ about $A = F \times AC$

$\quad\quad\quad\quad\quad\quad\quad\quad\quad = Fa\sin\theta$

### Method 2

Resolve $F$ into two components:

$\quad F\cos\theta$ in the direction $AB$

$\quad F\sin\theta$ perpendicular to $AB$

The first of these components is along a line through $A$, so its moment about $A$ is zero. The moment of $F$ is produced entirely by the second component.

Hence,

$\quad$ Moment of $F$ about $A = F\sin\theta \times a = Fa\sin\theta$

You should be able to use both methods, so that you can choose the best method to use in a given circumstance.

Find the total moment about the point $A$ of the forces shown in the diagram.

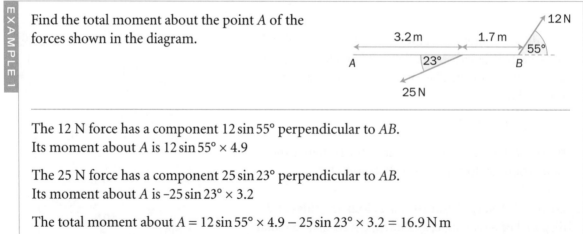

The 12 N force has a component $12\sin 55°$ perpendicular to $AB$.
Its moment about $A$ is $12\sin 55° \times 4.9$

The 25 N force has a component $25\sin 23°$ perpendicular to $AB$.
Its moment about $A$ is $-25\sin 23° \times 3.2$

The total moment about $A = 12\sin 55° \times 4.9 - 25\sin 23° \times 3.2 = 16.9\,\text{N m}$

**EXAMPLE 2**

The diagram shows a rectangular lamina *ABCD*.
A force of 20 N is applied at *C*, as shown.
Find the moment of this force about *A*.

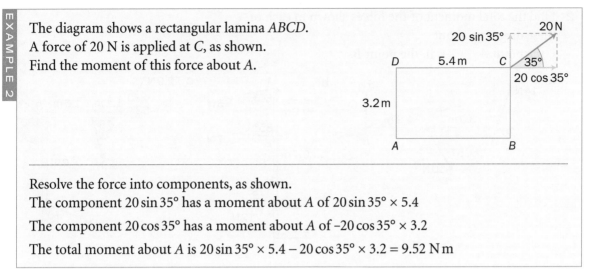

Resolve the force into components, as shown.

The component $20 \sin 35°$ has a moment about *A* of $20 \sin 35° \times 5.4$

The component $20 \cos 35°$ has a moment about *A* of $-20 \cos 35° \times 3.2$

The total moment about *A* is $20 \sin 35° \times 5.4 - 20 \cos 35° \times 3.2 = 9.52$ N m

Sometimes forces may be given using their vector components and
points may be given either as $(x, y)$ coordinates or as position vectors.

**EXAMPLE 3**

The force $\mathbf{F} = (5\mathbf{i} + 2\mathbf{j})$ N acts at the point *Q* with position vector $(4\mathbf{i} + 5\mathbf{j})$ m.
Find the moment of $\mathbf{F}$ about the point *P* with position vector $(\mathbf{i} + 3\mathbf{j})$ m.

From the diagram, you can see that

the 5 N component has a clockwise moment about
*P* of $-5 \times 2 = -10$ N m
the 2 N component has an anticlockwise moment about
*P* of $2 \times 3 = 6$ N m

Therefore
total moment about $P = -10 + 6 = -4$ N m

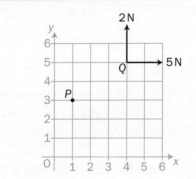

## Exercise 7.2

1 Find the moment of each of these forces about the point *A*,
indicating whether it is a positive or negative moment.

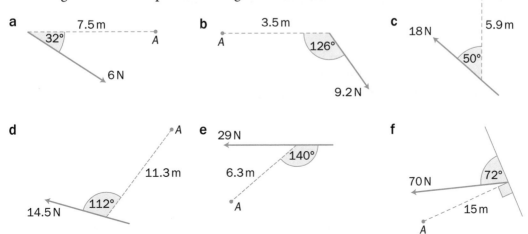

2 Find the total moment of the forces shown in each of these diagrams about

   **i** the point *A*        **ii** the point *B*.

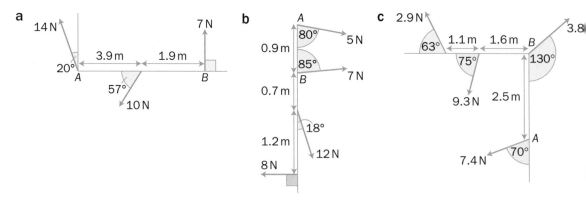

a

b

c

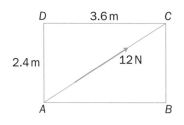

3 A force of 12 N acts along the diagonal *AC* of a rectangular lamina *ABCD*, as shown. Find the moment of the force about the point *B*.

4 A force **F** = (4**i** + 6**j**) N acts at the point *Q*, with position vector (4**i** + 7**j**) m. Find the moment of **F** about the point *P*, with position vector (2**i** + **j**) m.

5 A force **F** = (2**i** − 3**j**) N acts at the point *Q*, with position vector (2**i** + 4**j**) m. Find the moment of **F** about the point *P*, with position vector (**i** − 3**j**) m.

6 A force **F** = (−3**i** + 5**j**) N acts at the point *Q*, with position vector (−3**i** − 5**j**) m. Find the moment of **F** about the point *P*, with position vector (−5**i** + **j**) m.

7 Forces of 6**i** N, (2**i** + **j**) N and (**i** − 3**j**) N act respectively at the points with position vectors (3**i** + 2**j**) m, −4**j** m and (−2**i** + 5**j**) m. Find the total moment of the forces about

   **a**   the origin

   **b**   the point with position vector (**i** − **j**) m.

M1

8  The diagram shows a framework *ABCD* made of light rods each of length 2*a* m. Forces of 50 N, 80 N, 60 N and **P** act as shown.

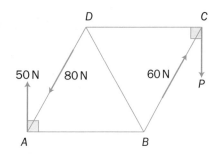

The total moment of the forces about the point *D* is $\frac{3}{4}$ of their total moment about the point *B*.

a  Show that **P** = 20 N.

b  Find the total moment of the forces about the point *A*.

9  Triangle *ABC* has vertices *A*(2, 3), *B*(6, 3) and *C*(6, 6). Forces of magnitude 12 N, 15 N and 30 N act along *AB*, *CB* and *AC* respectively, with directions given by the order of the letters.

a  Express each force in the form $(x\mathbf{i} + y\mathbf{j})$ N.

b  Find the total moment of the forces about

i  point *A*      ii  point *B*      iii  the origin.

10  The diagram shows a light rectangular framework, 4 m long by 3 m wide. The framework is acted on by four forces, as shown.

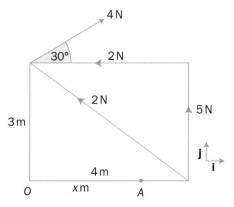

a  Express the forces in component form, using the **i**- and **j**-directions indicated. Hence find the resultant of the forces.

b  Find the total moment of the forces about the point *O*.

Varignon's theorem states that the sum of the moments of a system of forces about a point is the same as the moment of their resultant about that point.

c  The resultant of the forces acts through the point *A*, as shown, at a distance *x* m from *O*. Use Varignon's theorem to find the value of *x*.

11  *ABCD* is a rhombus, with centre *O*. *OB* = *a*. Forces of *P*, *P*, *Q* and *Q* act along *AB*, *BC*, *CD* and *DA*, as shown, where *P* > *Q*.

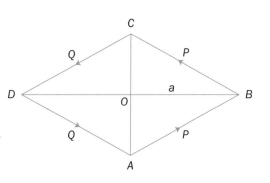

Use the methods from question **10** to show that the perpendicular distance of the line of action

of the resultant of these forces from *O* is $\dfrac{a(P + Q)}{(P - O)}$.

M1

If forces act on a particle, the particle accelerates unless the forces are in equilibrium. They are in equilibrium if their resultant is zero.

Forces acting on a particle are **concurrent** – they pass through a single point.

When forces act on a larger object, the forces may not be concurrent. In this case, you need more information to decide whether they are in equilibrium.

The simplest example of non-concurrent forces is when they are parallel. Such forces may have a resultant force, but they may also have a turning effect – a total moment – about some point.

This system of forces has a resultant force.
     Resultant force = 10 + 8 − 12 = 6 N

There is a turning effect about the point $A$.
     Total moment about $A = 8 \times 4 - 12 \times 3 = -4 \, \text{N m}$

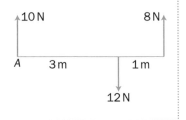

In general, an object acted upon by a system of parallel forces may have both linear acceleration and rotational acceleration. Both of these accelerations must be zero for the object to be in equilibrium.

A system of parallel forces is in equilibrium if
- the resultant of the forces is zero    **and**
- the total moment of the forces about any point is zero.

**Both** conditions must be met if the forces are to be in equilibrium.

For this system of forces
     Resultant force = 3 + 9 − 12 = 0 N

The turning effect about the point $A$
Total moment about $A = 9 \times 4 - 12 \times 3$
                 $= 0 \, \text{N m}$

The turning effect about the point $B$
Total moment about $B = 12 \times 1 - 3 \times 4$
                 $= 0 \, \text{N m}$

This system of forces is in equilibrium.

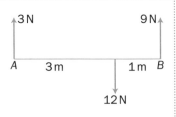

If the resultant is zero, you only need to show a zero moment about one point ($A$, say) to prove that the system is in equilibrium. You can examine the moment about a second point ($B$, say) as a check.

MI

A system of forces may have a zero resultant, yet not be in equilibrium.

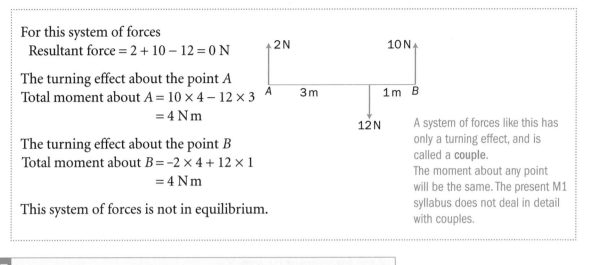

For this system of forces
   Resultant force = $2 + 10 - 12 = 0$ N

The turning effect about the point $A$
Total moment about $A = 10 \times 4 - 12 \times 3$
                            $= 4$ N m

The turning effect about the point $B$
Total moment about $B = -2 \times 4 + 12 \times 1$
                            $= 4$ N m

This system of forces is not in equilibrium.

A system of forces like this has only a turning effect, and is called a **couple**.
The moment about any point will be the same. The present M1 syllabus does not deal in detail with couples.

**EXAMPLE 1**

A uniform beam $AB$ of mass 10 kg and length 4 m rests in a horizontal position on a single support at $C$, 1 metre from $A$. The other end of the beam is supported by a vertical string, as shown.

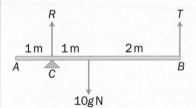

Find the reaction, $R$, at the support and the tension, $T$, in the string.

For a system of parallel forces to be in equilibrium
 o if you resolve in the direction of the forces, the resultant is zero
 o if you take moments about any point, the total moment is zero.

Resolve vertically:
$$R + T - 10g = 0 \qquad [1]$$

Take moments about $C$:
$$3T - 10g = 0 \qquad [2]$$

From [2],
$$T = 3\tfrac{1}{3}g = 32.7 \text{ N}$$

Substitute into [1]:
$$R = 6\tfrac{2}{3}g = 65.3 \text{ N}$$

It was better to take moments about $C$ rather than $A$ because $R$ has zero moment about $C$ and so only one unknown ($T$) appears in Equation [2].
Similarly, instead of resolving vertically, you could take moments about $B$ to find the value of $R$ directly.

EXAMPLE 2

The diagram shows a light rod $AB$ of length 3 m.
The point $C$ divides $AB$ in the ratio 2:1, and forces of
6 N, 7 N and 8 N act at $A$, $B$ and $C$ as shown.

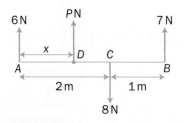

A fourth force, $P$ N, is applied to a point $D$ on the rod, where
$AD = x$ m, so that the system is then in equilibrium.
Find the values of $P$ and $x$.

For equilibrium, the resultant of the forces must be zero.

Resolve vertically:
$$P + 6 + 7 - 8 = 0 \quad \Rightarrow \quad P = -5$$

So you must apply a downward force of 5 N at $D$ to
achieve equilibrium.

For equilibrium, the total moment about any point
must be zero.

Take moments about $A$:
$$Px + 7 \times 3 - 8 \times 2 = 0$$
Substitute $P = -5$:
$$-5x + 21 - 16 = 0 \quad \Rightarrow \quad x = 1$$
So you apply the force $P$ 1 m from $A$.

EXAMPLE 3

A uniform beam $AB$, of mass 20 kg and length 3 m, rests
horizontally on two supports at $A$ and at $C$, where $AC = 2$ m.
Find the reactions at both supports.

If a beam is described as **uniform**
it means its weight acts through
the centre of the beam, which is
its centre of gravity. If a beam is
non-uniform, either you will be
told where its centre of gravity is
located or you will have to find
where it is. Example 4 involves a
non-uniform beam.

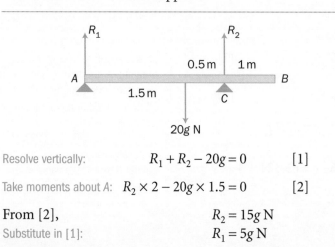

Resolve vertically: $\qquad R_1 + R_2 - 20g = 0 \qquad$ [1]

Take moments about $A$: $\quad R_2 \times 2 - 20g \times 1.5 = 0 \qquad$ [2]

From [2], $\qquad\qquad\qquad\qquad R_2 = 15g$ N
Substitute in [1]: $\qquad\qquad\qquad R_1 = 5g$ N

**EXAMPLE 4**

A non-uniform beam $AB$, of length 4 m and weight 500 N, rests horizontally on supports at $A$ and $B$. The reaction at $B$ is 140 N more than the reaction at $A$. Find the position of the centre of gravity of the beam.

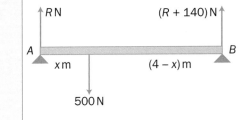

Resolve vertically:
$$R + R + 140 - 500 = 0$$
$$\Rightarrow \quad R = 180 \text{ N}$$

Take moments about $A$: $(R + 140) \times 4 - 500x = 0$

Substitute $R = 180$:
$$1280 - 500x = 0$$
$$\Rightarrow \quad x = 2.56$$

So the centre of gravity is 2.56 m from $A$.

In some situations an object may rest in equilibrium but be in a critical state, so that you could make it move by a very small change in the magnitude or position of a force.

**EXAMPLE 5**

The diagram shows a uniform beam $AB$, of weight 360 N and length 10 m. It is supported horizontally at $A$ and at $C$, where $AC = 6$ m. An object of weight 120 N is attached to the beam between $B$ and $C$ at a distance of $x$ m from $C$.

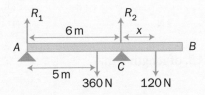

**a** Find the reactions at $A$ and $C$ in terms of $x$.

**b** Explain what happens as $x$ increases.

**a** Take moments about $C$:
$$360 \times 1 - 120x - R_1 \times 6 = 0$$
$$\Rightarrow \quad R_1 = (60 - 20x) \text{ N}$$

Resolve vertically:
$$R_1 + R_2 - 360 - 120 = 0$$
$$\Rightarrow \quad R_2 = 480 - R_1$$
$$= (420 + 20x) \text{ N}$$

**b** As $x$ increases, $R_1$ decreases and $R_2$ increases.
The system reaches a critical stage when $x = 3$, at which point $R_1 = 0$ N and $R_2 = 480$ N.
The system is then **on the point of tilting** about $C$, and the slightest increase in $x$ would cause the beam to rotate clockwise about $C$.

## Exercise 7.3

1  In each of these diagrams, a number of forces act perpendicular to a light rod, $AB$, of length 4 m. Determine in each case whether the rod is in equilibrium or not, giving reasons for your answer.

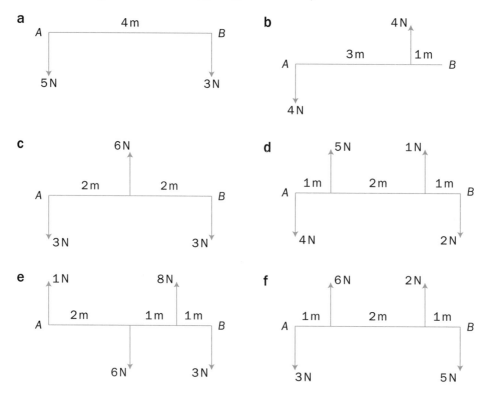

2  The diagram shows forces acting on a rod, $AB$, of length 5 m.

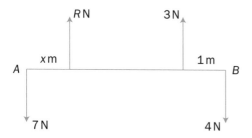

Find the values of $R$ and $x$ if the rod is in equilibrium.

3  A uniform beam, $AB$, of length 4 m and mass 50 kg rests on supports at $A$ and $B$. Objects of mass 20 kg and 40 kg hang from the beam at $C$ and $D$ respectively, where $AC = 1.4$ m and $AD = 3.2$ m.
Find the reactions at the two supports.

4  A light rod, *AB*, of length 2 m is suspended in a horizontal position by two vertical strings at *A* and *B*.

   a  An object of weight 200 N is suspended from the rod at *C*, where *AC* = 0.5 m. Calculate the tensions in the strings.

   b  The strings have a breaking strain of 180 N. The object is gradually moved along the rod towards *A*. How close can it be to *A* before the string breaks?

5  A light rod, *AB*, of length 5 m, rests horizontally on supports at *A* and at *C*, where *AC* = 4 m. A particle of mass 20 kg is attached to the rod at *D*, where *AD* = 1 m. A second particle of mass *m* is attached to the rod at *B*.

   a  Find the reactions at both supports in terms of *m*.

   b  Find the range of values of *m* for which the rod will remain in equilibrium.

6  The diagram shows a non-uniform beam, *AB*, of length 6 m and weight 500 N. It rests on supports at *A* and at its mid-point *C*. The centre of gravity of the beam is at *D*, between *A* and *C*. The reaction of the support at *C* is twice that at *A*.

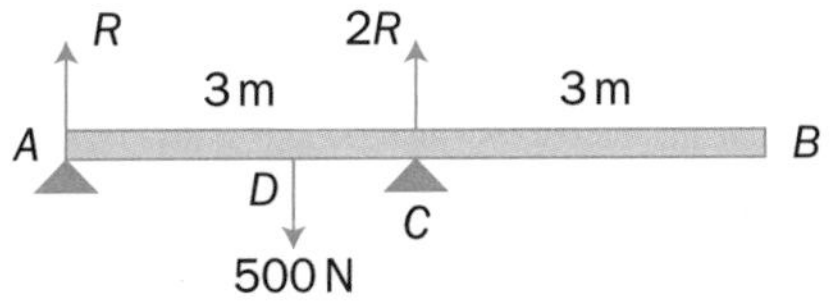

   a  Find the distance *CD*.

   b  A child of weight 200 N sits on the beam at a point between *B* and *C*. How far is the child from *C* if the beam is on the point of tilting?

7  A uniform beam, *AB*, of length 6 m and mass 100 kg, rests horizontally on supports at *C* and *D*, where *AC* = 2 m and *AD* = 5 m. A particle of mass 80 kg is attached to the beam at *A*, and a second particle, of mass *m* kg, is attached to the beam at *B*. Find the range of values of *m* for which the beam will remain in equilibrium.

M1

8 A non-uniform beam, *AB*, of weight 200 N and length 5 m, rests horizontally on supports at *C* and *D*, where *AC* = 1 m and *AD* = 3 m. If particles of weight 5*W* and *W* are attached at *A* and *B* respectively, the beam is on the point of tilting about *C*. If particles of weight *W* and 3*W* are attached at *A* and *B* respectively, the beam is on the point of tilting about *D*.

   **a** Find the distance of the centre of gravity of the beam from *A*.

   **b** Find the value of *W*.

9 A uniform rod, *AB*, of length 12*a* and weight 4*W* is suspended in a horizontal position by strings attached at *C* and *D*, where *AC* = 3*a* and *BD* = 4*a*. The breaking strain of the string at *C* is 3*W* and the breaking strain of the string at *D* is 3.8*W*.
An object of weight *W* is attached to the beam at a distance *x* from *A*.
Find the range of values of *x* if neither string is to break.

10 A non-uniform beam, *AB*, of length *a*, has weight *W* which acts through its centre of gravity which is at a distance *b* from *A*. The beam is placed symmetrically on two supports a distance *c* apart. Find, in terms of *W*, *a*, *b* and *c*, the reactions at the two supports.

11 A uniform plank is 12 m long and has mass 100 kg. It is placed on horizontal ground at the edge of a cliff, with 4 m of the plank projecting over the edge of the cliff.

   **a** How far out from the cliff can a man of mass 75 kg safely walk?

   **b** The man wishes to walk to the end of the plank. What is the minimum mass he should place on the other end of the plank so that he may safely walk to the end?

12 A heavy beam, *AB*, rests on two supports at points *C* and *D*, where *CD* = *a*. An object of weight *W* rests on the beam. If the object is moved a distance *b* in the direction *DC*, show that, provided equilibrium is maintained, the reaction at *C* will be increased by $\dfrac{Wb}{a}$.

13 The diagram shows a crane, comprising a rectangular housing *ABDE* of length 5 m and a jib *DC* of length 6 m inclined at 40° to the horizontal. The housing rests on supports at *A* and *B*, as shown.

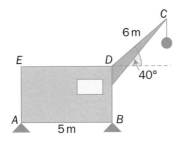

When there is no load at *C*, the reactions at *A* and *B* are 225 000 N and 275 000 N respectively.

a Find the horizontal distance from *A* to the centre of gravity of the crane.

b The safety regulations state that the load on the crane must not exceed 20% of the load which would cause the crane to tip over. Show that the mass of the maximum load is approximately 5 tonnes.

14 A non-uniform rod, *AB*, of length 2 m rests horizontally on supports *C* and *D*, where $AC = BD = 0.4$ m. If a particle of mass 8 kg is placed at *A*, the rod is on the point of tilting about *C*. If instead a particle of mass 10 kg is placed at *B*, the rod is on the point of tilting about *D*. Find the mass of the rod and the distance of its centre of gravity from *A*.

15 A non-uniform rod, *AB*, of length 4 m has its centre of gravity a distance *y* m from *A*. The rod will balance in a horizontal position on a single support *x* m from *A* if a particle of weight $2W$ N is placed at *A*. If instead a particle of weight $3W$ N is placed at *B*, the rod will balance on a single support placed *x* m from *B*.

a Find *y* in terms of *x*.

b Explain why $1.6 < y < 2$

16 A non-uniform rod, *AB*, of length 5 m and mass *M* kg, is placed in a horizontal position on two supports *C* and *D*, where $AC = 1$ m and $AD = 2$ m. If equal particles of mass $m_1$ kg are placed at *A* and *B*, the rod is on the point of tilting about *C*. If instead equal particles of mass $m_2$ kg are placed at *A* and *B*, the rod is on the point of tilting about *D*. Find *M* in terms of $m_1$ and $m_2$.

1 The diagram shows a rectangle, *ABCD*, with *AB* = 8 m and
*BC* = 4 m. Forces of 5 N, 3 N, 4 N and 6 N act along *AB*, *BC*,
*AC* and *DB*, as shown.
Calculate the total moment of this system of forces about

**a** *O*          **b** *A*          **c** *B*

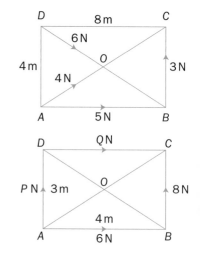

2 Forces of 6 N, 8 N, *P* N and *Q* N act along the sides *AB*, *BC*,
*DA* and *DC* of a rectangular lamina *ABCD*, where *AB* = 4 m and
*BC* = 3 m, as shown in the diagram. *O* is the centre of the lamina.

The moment of the forces about *A* is equal to that about *O*. If
the direction of the force *P* were reversed, the moment about
*O* would be twice that about *A*.
Calculate the values of *P* and *Q*.

3 The diagram shows a light rod, *AD*, of length
6 m. Points *B* and *C* lie on the rod so that
*AB* = *BC* = *CD*. Forces of 10 N, 6 N and 5 N act at
*B*, *C* and *D*, as shown.

Calculate the total moment of this system of
forces about

**a** *A*          **b** *C*

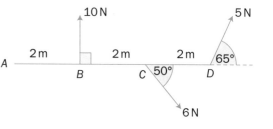

4 Forces of $(3\mathbf{i} + \mathbf{j})$ N and $(-2\mathbf{i} + 3\mathbf{j})$ N act at the points with
position vectors $(2\mathbf{i} + 3\mathbf{j})$ m and $(5\mathbf{i} + \mathbf{j})$ m respectively.
Calculate the total moment of these forces about

**a**   the origin

**b**   the point with position vector $(6\mathbf{i} + 3\mathbf{j})$ m.

5 A force of magnitude 30 N acts at the point with position vector
$(3\mathbf{i} + \mathbf{j})$ m in a direction inclined at 30° to the *x*-axis. A second
force, of magnitude 6 N, acts at the point with position vector
$(5\mathbf{i} - \mathbf{j})$ m in a direction inclined at 60° to the *x*-axis.
Find the total moment of these forces about the origin.

6 A uniform rod, *AB*, of mass 5 kg and length 4 m, is suspended
in a horizontal position by vertical strings attached to *A* and *B*.
A particle of mass 3 kg is attached to the rod at the point *C*,
where *AC* = 1 m. Calculate the tensions in the strings.

7 A uniform rod, *AB*, of length 2 m and weight 20 N, is suspended in a horizontal position by vertical strings at *A* and *B*. Each string has a breaking strain of 60 N. On what parts of the rod could you attach a particle of weight 70 N without breaking either string?

8 A uniform beam, of mass 40 kg and length 4 m, rests symmetrically on two supports which are 3 m apart. A particle of mass *m* kg is attached to one end of the rod.

   a   Find the reactions at the two supports if $m = 10$.

   b   Find the value of *m* if the beam is on the point of tilting about one of the supports.

9 Angela, of mass 40 kg, and Bill, of mass 60 kg, sit at the ends of a see-saw of length 3 m. Clare, of mass 30 kg, sits on the see-saw at a point so that the see-saw balances.

   a   Modelling the children as particles, find how far from the centre of the see-saw Clare must sit.

   b   Explain the significance of modelling the children as particles in part a.

10

A plank, *AB*, has length 4 m. It lies on a horizontal platform, with the end *A* lying on the platform and the end *B* projecting over the edge, as shown. The edge of the platform is at the point *C*.

Jack and Jill are experimenting with the plank. Jack has mass 40 kg and Jill has mass 25 kg. They discover that, if Jack stands at *B* and Jill stands at *A* and $BC = 1.6$ m, the plank is in equilibrium and on the point of tilting about *C*. By modelling the plank as a uniform rod, and Jack and Jill as particles,

   a   find the mass of the plank.

They now alter the position of the plank in relation to the platform so that, when Jill stands at *B* and Jack stands at *A*, the plank is again in equilibrium and on the point of tilting about *C*.

   b   Find the distance *BC* in this position.

   c   State how you have used the modelling assumptions that
      i    the plank is uniform
      ii   the plank is a rod
      iii  Jack and Jill are particles.

[(c) Edexcel Limited 2002]

# 7 Exit ⟹

## Summary

Refer to

- When a force **F** acts at a perpendicular distance $d$ from a point $A$, its turning effect about the point has magnitude $|\mathbf{F}| \times d$.
  - This turning effect is the moment of the force (the torque)
  - The unit of moment is the newton metre $(\text{N m})$
  - Anticlockwise moments are positive; clockwise moments are negative
  7.1

- For a force at an angle, $\theta$, as shown, moment about $A = Fa \sin \theta$
  7.2

- The moment of a force about a point is the sum of the moments of its components about that point
  7.2
- A system of forces is in equilibrium if
  - the resultant of the forces is zero **and**
  - the total moment of the forces about any point is zero.
  7.3

## Links

The movement of the human body relies on moments and forces. The joints act as pivots, and the bones act as levers.

Understanding moments and how they work in the body is particularly useful for sports professionals. Weight lifters develop their muscles by increasing the weight at the end of certain levers in their body. Analysing muscle moments (particularly in the ankle) can enable athletes (such as runners and gymnasts) to remain more balanced and so improve their performance.

M1

1 A particle, *P*, of mass 0.4 kg is moving under the action of a constant force **F** newtons. Initially the velocity of *P* is $(6\mathbf{i} - 27\mathbf{j})$ m s$^{-1}$ and 4 s later the velocity of *P* is $(-14\mathbf{i} + 21\mathbf{j})$ m s$^{-1}$.

   **a**    Find, in terms of **i** and **j**, the acceleration of *P*.

   **b**    Calculate the magnitude of **F**.                 [(c) Edexcel Limited 2003]

2 A sledge of mass 20 kg rests on a horizontal surface of ice. The coefficient of friction between the sledge and the ice is 0.1. A light rope attached to the front of the sledge is pulled and the sledge accelerates at 0.15 m s$^{-2}$. Find the tension in the rope if

   **a**    the rope is horizontal

   **b**    the rope is inclined at 25° to the horizontal.

3

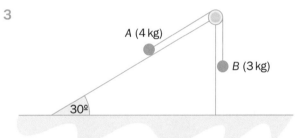

A particle, *A*, of mass 4 kg moves on the inclined face of a smooth wedge. This face is inclined at 30° to the horizontal. The wedge is fixed on horizontal ground. Particle *A* is connected to a particle *B*, of mass 3 kg, by a light inextensible string. The string passes over a small light smooth pulley which is fixed at the top of the plane. The section of the string from *A* to the pulley lies in a line of greatest slope of the wedge. The particle, *B*, hangs freely below the pulley, as shown in the diagram. The system is released from rest with the string taut. For the motion before *A* reaches the pulley and before *B* hits the ground, find

   **a**    the tension in the string

   **b**    the magnitude of the resultant force exerted by the string on the pulley.

   **c**    The string in this question is described as being 'light'.
      **i**    Write down what you understand by this description.
      **ii**   State how you used the fact that the string is light in your answer to part **a**.          [(c) Edexcel Limited 2004]

4 A stone, S, is sliding on ice. The stone is moving along a straight horizontal line ABC, where AB = 24 m and AC = 30 m. The stone is subject to a constant resistance to motion of magnitude 0.3 N. At A the speed of S is 20 m s$^{-1}$, and at B the speed of S is 16 m s$^{-1}$. Calculate

**a** the deceleration of S

**b** the speed of S at C.

**c** Show that the mass of S is 0.1 kg.

At C, the stone S hits a vertical wall, rebounds from the wall and then slides back along the line CA. The magnitude of the impulse of the wall on S is 2.4 Ns and the stone continues to move against a constant resistance of 0.3 N.

**d** Calculate the time between the instant that S rebounds from the wall and the instant that S comes to rest. [(c) Edexcel Limited 2005]

5    P (4 kg)                    Q (6 kg)        40 N

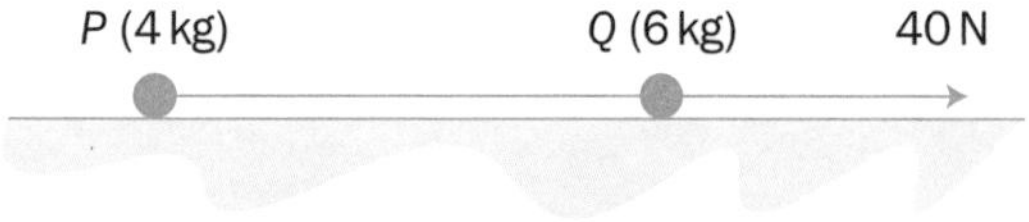

Two particles, P and Q, of mass 4 kg and 6 kg respectively, are joined by a light inextensible string. Initially the particles are at rest on a rough horizontal plane with the string taut. The coefficient of friction between each particle and the plane is $\frac{2}{7}$.

A constant force of magnitude 40 N is then applied to Q in the direction PQ, as shown in the diagram.

**a** Show that the acceleration of Q is 1.2 m s$^{-2}$.

**b** Calculate the tension in the string when the system is moving.

**c** State how you have used the information that the string is inextensible.

After the particles have been moving for 7 s, the string breaks. The particle Q remains under the action of the force of magnitude 40 N.

**d** Show that P continues to move for a further 3 seconds.

**e** Calculate the speed of Q at the instant when P comes to rest. [(c) Edexcel Limited 2004]

6  A

    B

A uniform rod, of mass 10 kg and length 4 m, is held in a horizontal position by two pegs, A and B, as shown in the diagram. The peg A is at the end of the rod and AB = 50 cm. Calculate the reactions at the two pegs.

**7**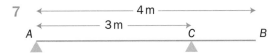

A uniform plank, *AB*, has mass 40 kg and length 4 m. It is supported in a horizontal position by two smooth pivots, one at the end *A*, the other at the point *C* of the plank where *AC* = 3 m, as shown in the diagram. A man of mass 80 kg stands on the plank which remains in equilibrium. The magnitudes of the reactions at the two pivots are each equal to *R* newtons. By modelling the plank as a rod and the man as a particle, find

**a** the value of *R*

**b** the distance of the man from *A*.

[(c) Edexcel Limited 2003]

**8**

A non-uniform rod, *AB*, has length 5 m and weight 200 N. The rod rests horizontally in equilibrium on two smooth supports *C* and *D*, where *AC* = 1.5 m and *DB* = 1 m, as shown in the diagram. The centre of mass of *AB* is *x* metres from *A*. A particle of weight *W* newtons is placed on the rod at *A*. The rod remains in equilibrium and the magnitude of the reaction of *C* on the rod is 160 N.

**a** Show that $50x - W = 100$.

The particle is now removed from *A* and placed on the rod at *B*. The rod remains in equilibrium and the reaction of *C* on the rod now has magnitude 50 N.

**b** Obtain another equation connecting *W* and *x*.

**c** Calculate the value of *x* and the value of *W*.

[(c) Edexcel Limited 2003]

**9** A particle, *A*, of mass 10 kg travelling at 2.4 m s⁻¹, collides with a second particle, *B*, of mass 5 kg, which is at rest. After the collision, the particles are travelling in the same direction, with the speed of *B* 0.9 m s⁻¹ greater than the speed of *A*.

**a** Find the speeds of the two particles after the collision.

After travelling a distance of 11 m at constant speed, *B* collides with a particle *C*, of mass 5 kg, which is at rest. *B* and *C* coalesce and move on together. Particle *A* now catches up with *B* and *C* and collides with them.

**b** Find the time which elapses between the second and third collisions.

M1

10 Arnold, whose mass is 40 kg, sits on a sledge of mass 10 kg at the top of a 20° slope of length 100 m. The coefficient of friction between the sledge and the snow is 0.1. The sledge starts from rest and accelerates down the slope.

    **a**   Calculate the magnitude of the acceleration.

Halfway down the slope the sledge collides with Beatrice, who is standing on the slope. Beatrice has mass 30 kg. She falls onto the sledge with Arnold, which then carries on down the slope.

    **b**   Calculate the speed of the sledge immediately after the impact.

    **c**   Show that the sledge takes nearly 10 s to get from the top of the slope to the bottom of the slope.

11 A car of mass 800 kg is towing a trailer of mass 200 kg along a straight horizontal road. The resistances to motion are 600 N for the car and 200 N for the trailer.

    **a**   The car exerts a driving force of 1200 N. Calculate
       **i**   the acceleration produced
      **ii**   the tension in the rigid coupling between the car and the trailer.

    **b**   The car brakes uniformly to rest from a speed of 30 m s$^{-1}$ in a distance of 150 m.
       **i**   Calculate the braking force needed.
      **ii**   Calculate the force in the rigid coupling.
     **iii**   Say, with reasons, whether the force in the coupling is a tension or a thrust.

12 A truck, *A*, of mass 3 tonnes moves on straight horizontal rails. It collides with a truck, *B*, of mass 1 tonne. Immediately before the collision the speed of *A* is 3 m s$^{-1}$, the speed of *B* is 4 m s$^{-1}$, and the trucks are moving towards each other. The trucks become coupled to form a single body, *C*, which continues to move on the rails.

    **a**   Find the speed and direction of *C* after the collision.

    **b**   Find the magnitude of the impulse exerted by *B* on *A* in the collision.

    **c**   State a modelling assumption which you have made about the trucks in your solution.

Immediately after the collision a constant braking force of magnitude 250 N is applied to *C*. It comes to rest in a distance *d* metres.

    **d**   Find the value of *d*.

[(c) Edexcel Limited 2003]

MI

# Answers

## Chapter 1

### Exercise 1.1

1  **a** Yes     **b** Probably not     **c** No
      **d** Yes     **e** Yes    **f** No
2  **a** Yes     **b** No     **c** No     **d** Yes
3  **a** 'Smooth' implies friction can be ignored.
      'Particle' means the tile has no significant
      size – it is a mass concentrated at a single point.
      The question probably also ignores air resistance.
   **b** Significant friction would mean the tile would
      leave the slope at a reduced speed. Significant size
      would mean that there might be rotational motion
      when the tile is in the air. Air resistance would
      reduce the acceleration of the tile both on the
      slope and in the air.
4  **a** No air resistance, balls are particles, strings are
      light and inextensible.
   **b** Air resistance would reduce the speed of impact. If
      the balls were of significant size, ball A would hit
      ball B before OA was vertical. A string of
      significant mass might affect the speed and effect
      of the impact. A string which could stretch would
      mean that OA and OB might not be equal at the
      moment of impact, so the impact would not be
      directly along the line of centres of the balls.

## Chapter 2

### Check in

1  3
2  103.5 units$^2$
3  88
4  **a** 5.5       **b** 6
5  **a** 2 or 3     **b** 3.30 or –0.303
6  **a** 0.350 kg    **b** 30 m s$^{-1}$

### Exercise 2.1

1  **a** 50 m, –40 m, 0 m    **b** 50 m, –90 m, 40 m
   **c** 180 m           **d** 0 m
2  **a** $81\frac{9}{11}$ km h$^{-1}$     **b** $22\frac{8}{11}$ m s$^{-1}$
3  3 km h$^{-1}$
4  **a** 20 km h$^{-1}$      **b** $5\frac{2}{9}$ m s$^{-1}$
5  **a** 7.58 km h$^{-1}$    **b** 12 km h$^{-1}$
6  **a** $22\frac{2}{9}$ m s$^{-1}$

   **b** The first stage takes 100 s and to average 40 m s$^{-1}$
      the whole journey would have to be done in
      this time.
7  **a** 3 m s$^{-1}$   **b** 4 min   **c** $3\frac{4}{7}$ m s$^{-1}$ **d** $\frac{5}{7}$ m s$^{-1}$
8  **a** 7.5 s    **b** velocity = –25 m s$^{-1}$, speed = 25 m s$^{-1}$
9  –1.25 m s$^{-2}$
10 **a** 0.77 m s$^{-1}$ **b** –0.21 m s$^{-1}$ **c** –0.25 m s$^{-2}$
11 **a** 96 km      **b** 25 m s$^{-1}$

12  $\frac{2V}{3}$ m s$^{-1}$

### Exercise 2.2

1  **a** Travel 10 km in 2 hours – velocity 5 km h$^{-1}$, rest
      for 1 hour – velocity 0 km h$^{-1}$, travel 10 km in 3
      hours – velocity $3\frac{1}{3}$ km h$^{-1}$.

   **b** Velocity constant during each stage, changes of
      velocity take place instantaneously, journey is a
      straight line.

2  **a**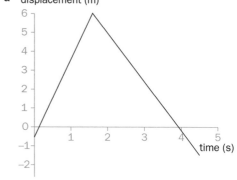

   **b** 3 m s$^{-1}$     **c** $-\frac{1}{3}$ m s$^{-1}$

3  **a**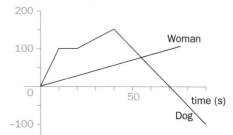

   **b** Pass after 54.7 s, 76.6 m from A
   **c** $4\frac{4}{9}$ m s$^{-1}$    **d** $-1\frac{1}{9}$ m s$^{-1}$

4  **a**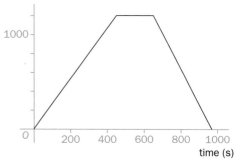

   Both speeds are constant. He starts and stops
   instantaneously.
   **b** 2.5 m s$^{-1}$    **c** 0 m s$^{-1}$

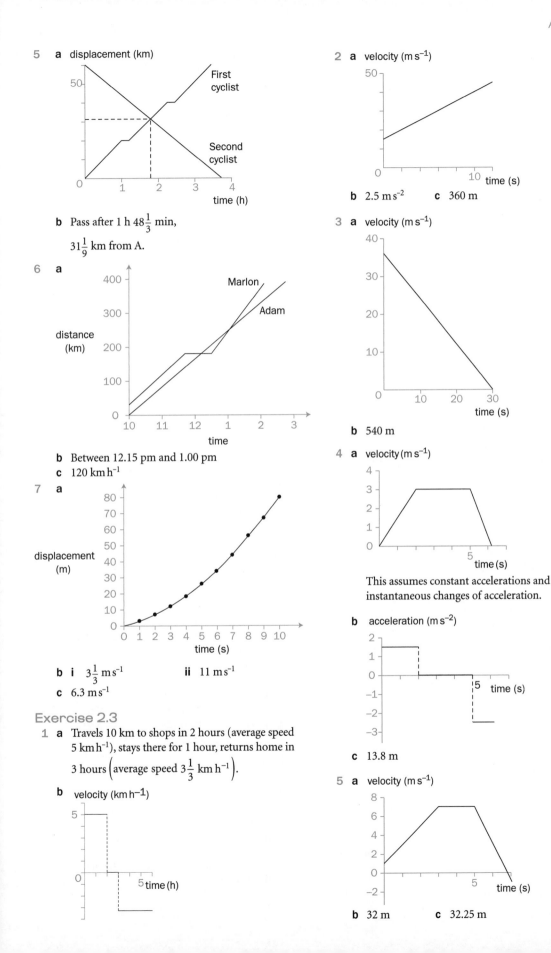

5 **a** displacement (km)

First cyclist

Second cyclist

**b** Pass after 1 h $48\frac{1}{3}$ min,

$31\frac{1}{9}$ km from A.

6 **a**

Marlon

Adam

distance (km)

time

**b** Between 12.15 pm and 1.00 pm

**c** 120 km h$^{-1}$

7 **a**

displacement (m)

time (s)

**b** **i** $3\frac{1}{3}$ m s$^{-1}$   **ii** 11 m s$^{-1}$

**c** 6.3 m s$^{-1}$

## Exercise 2.3

1 **a** Travels 10 km to shops in 2 hours (average speed 5 km h$^{-1}$), stays there for 1 hour, returns home in 3 hours $\left(\text{average speed } 3\frac{1}{3} \text{ km h}^{-1}\right)$.

**b** velocity (km h$^{-1}$)

time (h)

2 **a** velocity (m s$^{-1}$)

time (s)

**b** 2.5 m s$^{-2}$   **c** 360 m

3 **a** velocity (m s$^{-1}$)

time (s)

**b** 540 m

4 **a** velocity (m s$^{-1}$)

time (s)

This assumes constant accelerations and instantaneous changes of acceleration.

**b** acceleration (m s$^{-2}$)

time (s)

**c** 13.8 m

5 **a** velocity (m s$^{-1}$)

time (s)

**b** 32 m   **c** 32.25 m

M1

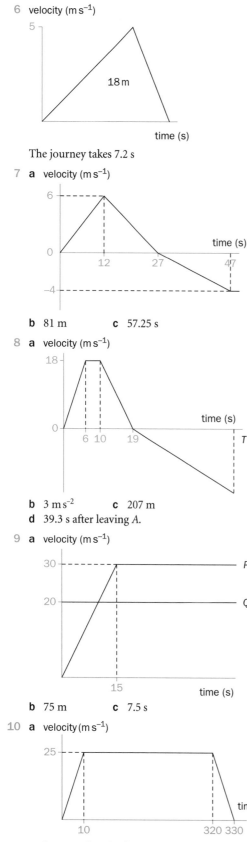

6   velocity (m s$^{-1}$)

The journey takes 7.2 s

7   **a**   velocity (m s$^{-1}$)

**b**   81 m          **c**   57.25 s

8   **a**   velocity (m s$^{-1}$)

**b**   3 m s$^{-2}$          **c**   207 m
**d**   39.3 s after leaving $A$.

9   **a**   velocity (m s$^{-1}$)

**b**   75 m          **c**   7.5 s

10   **a**   velocity (m s$^{-1}$)

Journey time 5 min 30 s

**b**   Maximum 7 min 33.6 s, when limit starts more than 230 m from Chulchit's home and ends more than 230 m from his work. Minimum 7 min 28.8 s when limit starts at Chulchit's home or ends at his work.

## Exercise 2.4

1   **a**   1350 m          **b**   90 m s$^{-1}$
2   **a**   10 m s$^{-2}$          **b**   125 m
3   **a**   3 m s$^{-2}$          **b**   30 m s$^{-1}$
4   4 s
5   7392 m
6   0.139 m s$^{-2}$, –0.069 m s$^{-2}$
7   **a**   $23\frac{5}{6}$ m          **b**   $3\frac{2}{3}$ s

   **c**   It avoids the need to consider the length of the car. In practice, the amount of level road between humps is the distance travelled by the car on the flat plus the distance between the front and back wheels of the car.
8   $a = 3.2$
9   35.5 s
10   15 m
11   **a**   –25 m          **b**   65 m
12   **a**   5 m s$^{-2}$          **b**   17.5 m s$^{-1}$
13   **a**   $-\frac{5}{6}$ m s$^{-2}$          **b**   1 s          **c**   $\frac{5}{12}$ m
14   3 m s$^{-1}$
15   **a**   60 s          **b**   1200 m
16   Henry stops 19.375 m behind Clare.
17   **a i**   5 s
      **ii**   8 m above P
      **b**   1.6 s after second ball starts, 17.28 m from the bottom.
19   $\frac{u^2}{2a}$
21   **a**   20 m s$^{-1}$, 10 m s$^{-1}$          **b**   750 m

## Exercise 2.5

1   **a**   3.19 s          **b**   31.3 m s$^{-1}$
2   **a**   11.5 m          **b**   1.53 s          **c**   3.06 s
3   **a**   19.8 m s$^{-1}$          **b**   4.04 s          **c**   19.8 m s$^{-1}$
4   **a**   34.7 m s$^{-1}$          **b**   4.05 s
5   **a**   3.88 m          **b**   1.52 s
6   45 m
7   **a**   Time is 4 s          **b**   $8g$ m
8   **a**   80 m          **b**   40 m s$^{-1}$          **c**   9.83 s
9   **a**   45.9 m          **b**   6.12 s          **c**   3.60 s
10   $\frac{5}{16}$ m above top of window
11   $V = h - 5.5g$
12   $\frac{u}{g} - \frac{T}{2}$
13   **a i**   $\frac{u}{g}$

      **ii**   $\frac{u^2}{2g}$

      **b i**   The velocity at $B$ is negative, –4.9 m s$^{-1}$
      **ii**   24.5 m s$^{-1}$
14   3 : 1

## Review 2

**1 a** velocity (m s⁻¹)

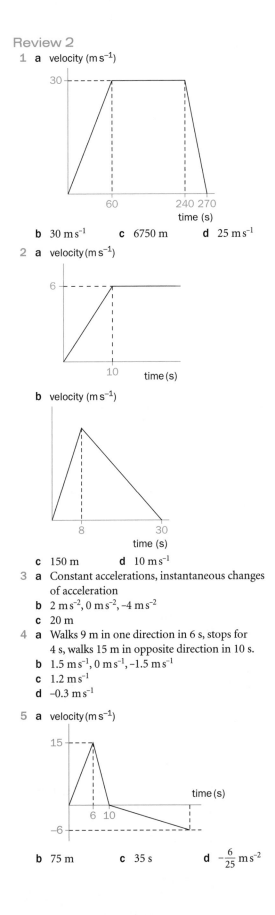

**b** 30 m s⁻¹     **c** 6750 m     **d** 25 m s⁻¹

**2 a** velocity (m s⁻¹)

**b** velocity (m s⁻¹)

**c** 150 m     **d** 10 m s⁻¹

**3 a** Constant accelerations, instantaneous changes of acceleration
   **b** 2 m s⁻², 0 m s⁻², –4 m s⁻²
   **c** 20 m

**4 a** Walks 9 m in one direction in 6 s, stops for 4 s, walks 15 m in opposite direction in 10 s.
   **b** 1.5 m s⁻¹, 0 m s⁻¹, –1.5 m s⁻¹
   **c** 1.2 m s⁻¹
   **d** –0.3 m s⁻¹

**5 a** velocity (m s⁻¹)

**b** 75 m     **c** 35 s     **d** $-\frac{6}{25}$ m s⁻²

**e** acceleration (m s⁻²)

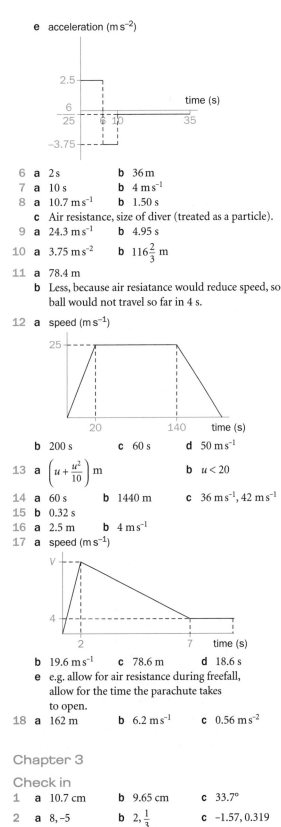

**6 a** 2 s            **b** 36 m
**7 a** 10 s           **b** 4 m s⁻¹
**8 a** 10.7 m s⁻¹     **b** 1.50 s
   **c** Air resistance, size of diver (treated as a particle).
**9 a** 24.3 m s⁻¹     **b** 4.95 s
**10 a** 3.75 m s⁻²    **b** $116\frac{2}{3}$ m

**11 a** 78.4 m
   **b** Less, because air resiatance would reduce speed, so ball would not travel so far in 4 s.

**12 a** speed (m s⁻¹)

   **b** 200 s        **c** 60 s        **d** 50 m s⁻¹
**13 a** $\left(u + \frac{u^2}{10}\right)$ m           **b** $u < 20$

**14 a** 60 s         **b** 1440 m      **c** 36 m s⁻¹, 42 m s⁻¹
**15 b** 0.32 s
**16 a** 2.5 m        **b** 4 m s⁻¹
**17 a** speed (m s⁻¹)

   **b** 19.6 m s⁻¹    **c** 78.6 m      **d** 18.6 s
   **e** e.g. allow for air resistance during freefall, allow for the time the parachute takes to open.
**18 a** 162 m        **b** 6.2 m s⁻¹   **c** 0.56 m s⁻²

## Chapter 3

### Check in

**1 a** 10.7 cm       **b** 9.65 cm     **c** 33.7°
**2 a** 8, –5         **b** 2, $\frac{1}{3}$     **c** –1.57, 0.319
**3 a** 350°          **b** 6.06 km

M1

## Exercise 3.1

1 a i $p + \frac{1}{2}q$  ii $\frac{1}{2}q - p$  iii $q - p$

2 a 2q  b p + q  c q − 2p
  d 2q − 2p  e p − 2q  f −p − q

3 a 2q  b p + q  c −q

  d 2q − p  e $\frac{1}{2}p + 1\frac{1}{2}q$  f $\frac{1}{2}p - 1\frac{1}{2}q$

4 a i q − p  ii $\frac{1}{2}q - \frac{1}{2}p$

  b BC is parallel to EF and twice as long.
6 a 7.02 km, 098.8°  b 23.4 km, 217°
  c 23.5 km h⁻¹, 291°  d 529 N, 348°

7 a $\overrightarrow{AB} = \overrightarrow{OB} - \overrightarrow{OA}$  b i 20.6 km, 141°
  ii 61.8 km, 141°  iii 20.6t km, 141°
  c 5.82 h (5 h 49 min)
  d 10.49 h (10 h 29 min)
8 a 5.39 m s⁻¹, 80 m downstream from B
  b Steer upstream at 23.6° to AB, 4.58 m s⁻¹
9 a 408 km h⁻¹, 078.7°  b 101.5°, 392 km h⁻¹
10 6 km h⁻¹, 10.4 km h⁻¹
11 The race is a dead heat – both boats take 395 s
12 $\sqrt{v^2 - u^2} : v$
13 a Upstream at 41.8° to AB  b 40.2 s
   c 60 m  d 5.85 m s⁻¹

## Exercise 3.2

1 a 2j  b −2i − 2j  c −6i + 7j
  d 16i + j  e $\sqrt{5}$  f $2\sqrt{5}$
2 a u = −10, v = −1  b −4.5
3 a −12i + 16j  b −0.6i + 0.8j
4 a $x = \frac{2}{3}, y = 3\frac{1}{3}$

  b x = 2, y = 1 or x = −2, y = −1
5 x = 12, y = 15
6 a 5.80i + 3.91j  b 2.00i + 9.39j
  c 8.9i + 8.0j  d 4j
  e −3.64i + 7.46j  f −10.1i + 21.8j
  g −3.56i − 5.08j  h 12.5i − 6.4j
  i −2.06i − 2.83j
7 a 5.39, 21.8°  b 5, −90°
  c 3.61, 124°  d 5.83, −59.0°
  e 7.81, −140°  f 2, 180°
8 a 19.0i + 6.18j, 11.4i + 25.6j
  b 30.4i + 31.8j  c 44.0 km, 043.8°
9 a −34.5i + 6.08j, 25i + 43.3j
  b $\overrightarrow{AB} = \overrightarrow{OB} - \overrightarrow{OA} = 59.5i + 37.2j$
  c 70.2 km, 058°
10 a wind 35.4i + 35.4j, resultant velocity 282i − 103j,
    still air velocity 282.5 km h⁻¹ on a bearing of 119°
   b 273 km h⁻¹ on a bearing of 129°

## Exercise 3.3

1 a 7.21 m  b 33.7°
2 a (6i + 9j) m  b 8.54 m, 69.4° to x-direction
3 a 5.83 m s⁻¹  b 31.0°
4 a 11.4 m s⁻¹  b 37.9° to x-direction
5 a ((3t + 6)i − (2t + 1)j) m s⁻¹
  b It would require t < 0  c t = 2

6 Collide when t = 2
7 p = 7, q = 6
8 a 2.24 m s⁻¹
  b A at 63.4°, B at 18.4° to x-direction, angle
    between = 45°
  c $\overrightarrow{OA} = (ti + 2tj)$ m, $\overrightarrow{OB} = (3ti + tj)$ m,
    $\overrightarrow{AB} = (2ti - tj)$ m
  d 40.25 s
9 a (5ti + 10tj) km
  b ((5t − 30)i + (10t − 20)j) km
  c 2 hours  d 20 km  e 3.6 h
10 2.53 km after 24 min
11 a (1 + 3t)i + (3 + t)j, (4 + t)i + (2 + 2t)j
   b (3 − 2t)i + (−1 + t)j
   c $\overrightarrow{AB}$ can never be the zero vector
   d 3.6 m  e 4.5 m s⁻¹
12 a p = (3ti + (20 − 2t)j) km
     q = (6ti + 2tj) km
   b (3ti + (4t − 20)j) km
   d 3.2 h or 3 h 12 min

## Review 3

1 a 50 m  b 30° upstream from AB, 28.9 s
2 a 5.10  b 11.3°
3 a 6i + 10.4j, 15i  b 13.7 km, 139°
4 a 031.0°  b 9tj, (3t − 10)i + 5tj
  c 1520 hours  e 1424 hours
5 a (−2i + 4j) m s⁻²  b 35 m
6 a (15.6i − j) m s⁻¹
  b 15.6 m s⁻¹ on a bearing of 093.7°
  c Resultant velocity V given by any point P on
    the circle shown. V cannot be on a bearing
    of 240°

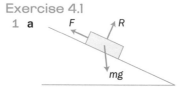

7 a 9.43 m s⁻¹  b (5t + 2)i + (8t + 1)j
  c t = 1.6 s  d 4.25 m s⁻¹
  e e.g. Friction between the ball and the ground.

## Chapter 4

## Check in

1 x = 38.7°, y = 10.4 cm
2 x = 4, y = 2.5
3 a 4.74i + 8.39j  b 9.63, 60.5° to x-direction

## Exercise 4.1

1 a

**b**

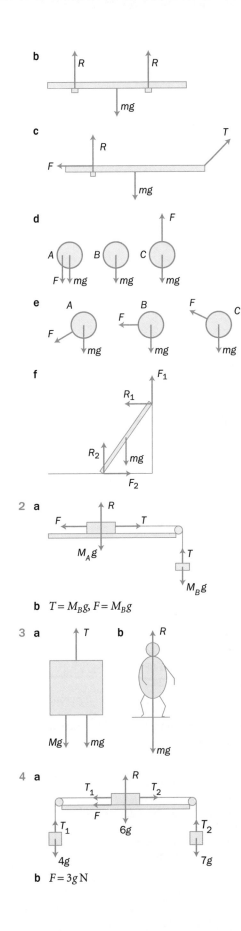

**c**

**d**

**e**

**f**

**2 a**

**b** $T = M_B g,\ F = M_B g$

**3 a** **b**

$Mg$ $mg$

$mg$

**4 a**

**b** $F = 3g\,\text{N}$

**5**

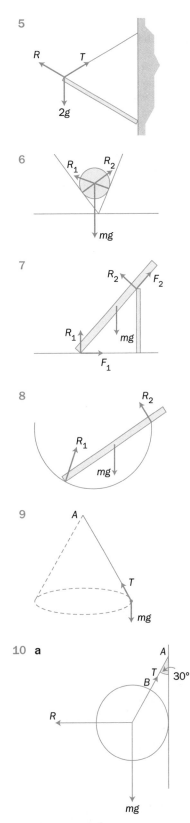

**6**

**7**

**8**

**9**

**10 a**

**b** The angle between the string and the wall might be different.

## Exercise 4.2

1  a  Resultant: $(4.31\mathbf{i} + 4.10\mathbf{j})$ N or 5.95 N at 43.5°
      to $x$-direction
      Equilibrant: $(-4.31\mathbf{i} - 4.10\mathbf{j})$ N or 5.95 N
      at $-136.5°$ to $x$-direction
   b  Resultant: $(-1.19\mathbf{i} + 2.98\mathbf{j})$ N or 3.21 N
      at 111.7° to $x$-direction
      Equilibrant: $(1.19\mathbf{i} - 2.98\mathbf{j})$ N or 3.21 N
      at $-68.3°$ to $x$-direction
   c  Resultant: $(-1.79P\mathbf{i} + 0.109P\mathbf{j})$ N or $1.80P$ N
      at 176.5° to $x$-direction
      Equilibrant: $(1.79P\mathbf{i} - 0.109P\mathbf{j})$ N or $1.80P$ N
      at $-3.5°$ to $x$-direction

2  a  $P = 9.18$ N, $Q = 9.66$ N
   b  $P = 7.66$ N, $Q = 6.73$ N
   c  $P = -3.23$ N, $Q = -12.1$ N

3  a  $P = 12.9$ N, $Q = 15.3$ N
   b  $P = 9.43$ N, $\theta = 148°$
   c  $P = 17.2$ N, $Q = 15.8$ N

4  a  $P = 2.54$ N, $Q = 5.44$ N
   b  $P = 17.3$ N, $Q = 24.5$ N
   c  $P = 17.3$ N, $Q = 24.0$ N

5  9.95 N, 15.2 N

6  $F = 30.2$ N, $R = 82.9$ N

7  a  $T = \dfrac{P\sqrt{3}}{2}$, $R = \dfrac{1}{2}(392 - P)$

   b  The block would lift off the ground.

8  a  60.1 N, 50.4 N
   b  No – cylinder has no tendency to rotate,
      so no friction forces needed.

9  $W = 91.2$ N, $\theta = 19.5°$

10  a  $P = 16.3$ N, $T = 42.5$ N
    b  15.1 N at right angles to the string, $T = 36.2$ N

11  a  17.9 N, 25.3 N
    b  $P = 5.60$ N, $T = 22.4$ N

12  a

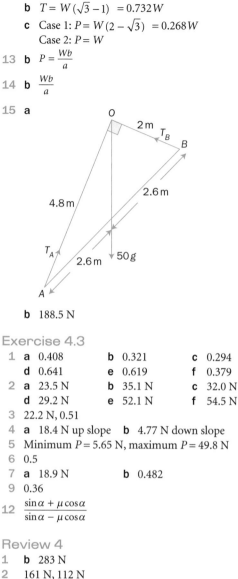

b  $T = W(\sqrt{3} - 1) = 0.732W$
c  Case 1: $P = W(2 - \sqrt{3}) = 0.268W$
   Case 2: $P = W$

13  b  $P = \dfrac{Wb}{a}$

14  b  $\dfrac{Wb}{a}$

15  a

b  188.5 N

## Exercise 4.3

1  a  0.408    b  0.321    c  0.294
   d  0.641    e  0.619    f  0.379

2  a  23.5 N    b  35.1 N    c  32.0 N
   d  29.2 N    e  52.1 N    f  54.5 N

3  22.2 N, 0.51

4  a  18.4 N up slope    b  4.77 N down slope

5  Minimum $P = 5.65$ N, maximum $P = 49.8$ N

6  0.5

7  a  18.9 N    b  0.482

9  0.36

12  $\dfrac{\sin\alpha + \mu\cos\alpha}{\sin\alpha - \mu\cos\alpha}$

## Review 4

1  b  283 N
2  161 N, 112 N
3  b  179 N
4  a  6.93 N    b  3.46 N
5  a  7.55 N    b  14.8°
6  a  52.6°    b  24.7 N
7  b  0.197
8  a  0.45    b  1.44 N
   c  No – reaction reduced, so available friction
      force reduced.

## Revision 1

1  a  31.1 N    b  10.1
2  a  13 m s$^{-1}$    b  1.2 m s$^{-2}$
3  a  $(5\cos\theta\,\mathbf{i} + 5\sin\theta\,\mathbf{j})$ N, $(4\sqrt{3}\mathbf{i} - 4\mathbf{j})$ N
   b  i  53.1°    ii  9.93 N
4  a  0.0382    b  37.0 N
5  a  86.6 N    b  30°

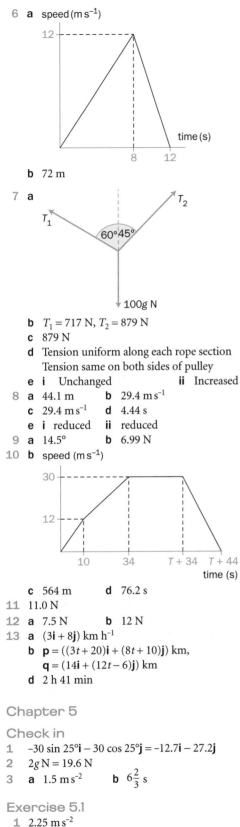

6 a speed (m s⁻¹)

b 72 m

7 a

b $T_1 = 717$ N, $T_2 = 879$ N
c 879 N
d Tension uniform along each rope section
  Tension same on both sides of pulley
e i Unchanged      ii Increased
8 a 44.1 m     b 29.4 m s⁻¹
  c 29.4 m s⁻¹   d 4.44 s
  e i reduced    ii reduced
9 a 14.5°     b 6.99 N
10 b speed (m s⁻¹)

  c 564 m     d 76.2 s
11 11.0 N
12 a 7.5 N     b 12 N
13 a $(3\mathbf{i} + 8\mathbf{j})$ km h⁻¹
  b $\mathbf{p} = ((3t + 20)\mathbf{i} + (8t + 10)\mathbf{j})$ km,
    $\mathbf{q} = (14\mathbf{i} + (12t - 6)\mathbf{j})$ km
  d 2 h 41 min

## Chapter 5

### Check in

1 $-30 \sin 25°\mathbf{i} - 30 \cos 25°\mathbf{j} = -12.7\mathbf{i} - 27.2\mathbf{j}$
2 $2g$ N = 19.6 N
3 a 1.5 m s⁻²     b $6\frac{2}{3}$ s

### Exercise 5.1

1 2.25 m s⁻²
2 52.5 N

3 $6\frac{2}{3}$ kg
4 $(3\mathbf{i} + 4.5\mathbf{j})$ m s⁻²
5 $(6\mathbf{i} - 15\mathbf{j})$ N
6 a 0.8 m s⁻²     b 400 kg     c 380 N
  d 1290 N     e −0.4 m s⁻²
7 $(0.75\mathbf{i} + 1.25\mathbf{j})$ m s⁻²
8 $(-\mathbf{i} - \mathbf{j})$ N
9 4.04 m s⁻²
10 0.306
11 702 N
12 a 9.50 m s⁻² at 51.2° to $x$-direction
  b 11.8 m s⁻² at 100.7° to $x$-direction
13 a 3.07 m s⁻²     b 59.2 N
14 a 1460 N     b 0.547 m s⁻²
15 a 539 N     b 15.6 m s⁻²
16 0.725 m s⁻² on bearing 073.3°
17 a 172.4 N, 127.9 N
  b 8.49 m s⁻² at 30° to the vertical
18 a 22.8 m s⁻¹     b 16.9°

### Exercise 5.2

1 a i 2.7 m s⁻² upwards ii 2.3 m s⁻² downwards
  b i 196 N     ii 196 N     iii 236 N
    iv $182\frac{2}{3}$ N     v 241 N
2 a 12 m s⁻¹     b 99 m
3 a 14.6 m
4 0.092
5 Stage 1: 18.7 N; stage 2: 14.7 N; stage 3: −9.3 N
6 10.8 m s⁻¹
7 a 4.59 m     b particle remains at rest
8 a 5.49 m s⁻¹     b 12.9 m s⁻¹

### Exercise 5.3

1 a 550 N     b 490 N     c 452.5 N
  d 390 N
2 a 13.1 kg     b 80 kg     c 92.2 kg
  d 73.5 kg     e 72.5 kg
3 1.96 m s⁻² upwards
4 3.27 m s⁻²
5 a 1030 N     b 824 N
6 a 39.2 N, 68.6 N     b 39.2 N, 68.6 N
  c 51.2 N, 89.6 N
7 a 90.4 N, 56.5 N, 56.5 N
  b 5.2 m s⁻², top string breaks
8 a 1 m s⁻², 1500 N     b 1200 N     c 136 N
9 a 45.8 N     b 29.5 N
10 a $\frac{g}{6}$ m s⁻²     b $5\frac{5}{6}g$ N     c $11\frac{2}{3}g$ N
11 $\frac{1}{5}g$ m s⁻², $\sqrt{\frac{8g}{5}}$ m s⁻¹
12 $\frac{g}{3}$ m s⁻², $1\frac{1}{3}gm$ N
13 3.92 m s⁻², 11.8 N
14 3.22 m s⁻², 32.9 N
15 5.88 m s⁻², 0.714 s
16 2.28 m s⁻², 22.5 N
17 a 4.2 m s⁻²     b 5.02 m s⁻¹     c 1.29 m

M1

**18 a** $0.891\,\mathrm{m\,s^{-2}}$ downwards

   **b** $17.8\,\mathrm{N}$

**20 a** $\frac{1}{2}a$             **b** $1.51\,\mathrm{m\,s^{-2}}$, $9.05\,\mathrm{N}$

## Review 5

**1 a** $3600\,\mathrm{N}$       **b** $600\,\mathrm{N}$

**2** $(2\mathbf{i} + 2\mathbf{j})\,\mathrm{m\,s^{-2}}$

**3 a** $6.88\,\mathrm{N}$       **b** $6.60\,\mathrm{N}$

**4 a** $47.2\,\mathrm{N}$       **b** $48.2\,\mathrm{m}$

**5 a** $(-10\mathbf{i} - 5\mathbf{j})\,\mathrm{N}$       **b** $(6\mathbf{i} + 19\mathbf{j})\,\mathrm{N}$

**6 a i** $1.2\,\mathrm{m\,s^{-2}}$       **ii** $320\,\mathrm{N}$

  **b i** $80\,\mathrm{N}$         **ii** $0\,\mathrm{N}$

**7 a** $1.4\,\mathrm{m\,s^{-2}}$       **b** $33.6\,\mathrm{N}$

**8 a** $0.98\,\mathrm{m\,s^{-2}}$       **b** $8.82m\,\mathrm{N}$

**9 a** $5.39\,\mathrm{m\,s^{-2}}$       **b** $(10\mathbf{i} - 17\mathbf{j})\,\mathrm{m\,s^{-1}}$

**10 a** $16.8\,\mathrm{N}$       **b** $1.2\,\mathrm{kg}$

**11 a** $0.24\,\mathrm{m\,s^{-2}}$       **b** $534\,\mathrm{N}$       **c** $54\,\mathrm{m}$

  **d** Increased by the vertical component of the tension (changes from $(900g - T\sin 15°)\,\mathrm{N}$ to $900g\,\mathrm{N}$)

**12 a** $1.2mg\,\mathrm{N}$       **b** $0.693$

  **c** $1.2mg\,\mathrm{N}$ vertically downwards.

**13 a** $0.4g - T = 0.08g$       **b** $3.14\,\mathrm{N}$

  **d** Tension is the same in both string sections

  **e** $\sqrt{\dfrac{2g}{5}} = 1.98\,\mathrm{m\,s^{-1}}$

## Chapter 6

### Check in

**1** $(15.3\mathbf{i} + 12.9\mathbf{j})\,\mathrm{m\,s^{-1}}$

**2** $a = 2.5$, $b = 6.5$

**3** $64\,\mathrm{m}$

**4** $-2.5\,\mathrm{N}$

### Exercise 6.1

**1** $7\,\mathrm{m\,s^{-1}}$

**2 a** $42\,\mathrm{N\,s}$       **b** $120\,\mathrm{N}$

**3** $7.2\,\mathrm{N\,s}$

**4** $50\,000\,\mathrm{N\,s}$, $16\,700\,\mathrm{N}$ (3 s.f.)

**5** $5\,\mathrm{m\,s^{-1}}$

**6** $(-12\mathbf{i} + 16\mathbf{j})\,\mathrm{N\,s}$

**7** $(4.4\mathbf{i} - 0.4\mathbf{j})\,\mathrm{m\,s^{-1}}$

**8** $4.5\,\mathrm{s}$

**9 a** $4\,\mathrm{kg}$       **b** $-40\,\mathrm{N\,s}$       **c** $40\,\mathrm{N}$

**10 a** $1.18\,\mathrm{m\,s^{-1}}$   **b** $11.8\,\mathrm{m\,s^{-2}}$

**11 a** $2u\mathbf{i}$, $u\sin 30\mathbf{i} + u\cos 30\mathbf{j}$       **b** $mu\sqrt{3}\,\mathrm{N\,s}$

  **c** Bearing of $300°$

### Exercise 6.2

**1** $3.97\,\mathrm{m\,s^{-1}}$

**2 a** $3.2\,\mathrm{m\,s^{-1}}$       **b** $0.8\,\mathrm{m\,s^{-1}}$

**3** $0.45\,\mathrm{kg}$

**4 a** $\dfrac{10v}{7}$       **b** $\dfrac{2v}{7}$

**5 a** $4\frac{2}{3}\,\mathrm{m\,s^{-1}}$

  **b** If $m > 4$, $A$'s velocity is greater than $B$'s, so $A$ would need to 'pass through' $B$

**6 a** $15\,\mathrm{m\,s^{-1}}$       **b** $6000\,\mathrm{N}$

---

**7** $717\,\mathrm{m\,s^{-1}}$

**8** $3.6\,\mathrm{m\,s^{-1}}$

**9** $2\,\mathrm{m\,s^{-1}}$

**10** $3600\,\mathrm{N}$

**11** $(3.2\mathbf{i} + 3.2\mathbf{j})\,\mathrm{m\,s^{-1}}$

**12** $-8\mathbf{j}\,\mathrm{m\,s^{-1}}$

**13** $(2.5\mathbf{i} + 0.5\mathbf{j})\,\mathrm{m\,s^{-1}}$

**14** $m = 2$, $a = -5$

**15** $3\frac{1}{3}\,\mathrm{m\,s^{-1}}$

**16 a** $\dfrac{v}{n}\,\mathrm{m\,s^{-1}}$       **b** $\dfrac{dn(n-1)}{2v}\,\mathrm{s}$

**17** $\dfrac{9}{14}\,\mathrm{m\,s^{-1}}$

**18** $1.48\,\mathrm{m\,s^{-1}}$

**19** $1.53\,\mathrm{m\,s^{-1}}$, $0.6\,\mathrm{m}$

## Review 6

**1 a** $8.4\,\mathrm{N\,s}$       **b** $42\,\mathrm{N}$

**2** $(6\mathbf{i} + 6\mathbf{j})\,\mathrm{N\,s}$

**3 a** Initial: $(1.93\mathbf{i} - 2.30\mathbf{j})\,\mathrm{m\,s^{-1}}$; final: $(1.73\mathbf{i} + \mathbf{j})\,\mathrm{m\,s^{-1}}$

  **b** $(-0.0294\mathbf{i} + 0.495\mathbf{j})\,\mathrm{N\,s}$

**4** $3.64\,\mathrm{N\,s}$

**5 a** $1\,\mathrm{m\,s^{-1}}$       **b** $6\,\mathrm{m\,s^{-1}}$

**6 a** $3\,\mathrm{m\,s^{-1}}$       **b** $4.76\,\mathrm{m\,s^{-1}}$

**7** $2\,\mathrm{m\,s^{-1}}$

**8** $\dfrac{u}{4}$, $\dfrac{u}{2}$

**9 a** $0.36\,\mathrm{N\,s}$   **b** $2.7\,\mathrm{m\,s^{-1}}$   **c** $0.216\,\mathrm{N\,s}$

**10 a** $2.4\,\mathrm{m\,s^{-1}}$   **b** West   **c** $3000\,\mathrm{kg}$

**11 b** $20\,\mathrm{m\,s^{-1}}$

**12 a** $15\,\mathrm{m\,s^{-1}}$   **b** $960\,\mathrm{N}$

**13 a** $2.25\,\mathrm{m\,s^{-1}}$   **b** $1.5\,\mathrm{N\,s}$

**14 a** $3\,\mathrm{m\,s^{-1}}$   **b i** $3.6\,\mathrm{kg}$   **ii** $18\,\mathrm{N\,s}$

## Chapter 7

### Check in

**1** $(4.10\mathbf{i} + 11.3\mathbf{j})\,\mathrm{N}$, $(-5.74\mathbf{i} + 8.19\mathbf{j})\,\mathrm{N}$

**2 a** $31.0°$       **b** $2.57\,\mathrm{m}$

### Exercise 7.1

**1 a** $+56\,\mathrm{N\,m}$   **b** $-87.5\,\mathrm{N\,m}$   **c** $-100.8\,\mathrm{N\,m}$

  **d** $-31.5\,\mathrm{N\,m}$   **e** $+451.2\,\mathrm{N\,m}$   **f** $+720\,\mathrm{N\,m}$

**2 a i** $+19\,\mathrm{N\,m}$   **ii** $-35.6\,\mathrm{N\,m}$

  **b i** $-18\,\mathrm{N\,m}$   **ii** $-24.4\,\mathrm{N\,m}$

  **c i** $-9.39\,\mathrm{N\,m}$   **ii** $-12.81\,\mathrm{N\,m}$

**3 a** $+14.4\,\mathrm{N\,m}$   **b** $-9.6\,\mathrm{N\,m}$   **c** $+2.4\,\mathrm{N\,m}$

  **d** $-4.8\,\mathrm{N\,m}$

**4 a** $1.1\,\mathrm{m}$   **b** $0.6\,\mathrm{m}$   **c** $0.8\,\mathrm{m}$

**5 a** $30\,\mathrm{N}$

**6** $P = 6$, $Q = 8$

### Exercise 7.2

**1 a** $+23.8\,\mathrm{N\,m}$   **b** $-26.1\,\mathrm{N\,m}$   **c** $-81.4\,\mathrm{N\,m}$

  **d** $-152\,\mathrm{N\,m}$   **e** $+117\,\mathrm{N\,m}$   **f** $+324\,\mathrm{N\,m}$

**2 a i** $+7.89\,\mathrm{N\,m}$   **ii** $-60.4\,\mathrm{N\,m}$

  **b i** $-10.2\,\mathrm{N\,m}$   **ii** $-17.0\,\mathrm{N\,m}$

  **c i** $+9.43\,\mathrm{N\,m}$   **ii** $-9.99\,\mathrm{N\,m}$

**3** $-24.0\,\mathrm{N\,m}$

**4** $-12\,\mathrm{N\,m}$

**5** $-17\,\mathrm{N\,m}$

6   $-8\,\mathrm{N\,m}$

7  **a**  $-3\,\mathrm{N\,m}$       **b**  $-10\,\mathrm{N\,m}$

8  **b**  $43.9a\,\mathrm{N\,m}$

9  **a**  $12\mathbf{i}\,\mathrm{N},\,-15\mathbf{j}\,\mathrm{N},\,(24\mathbf{i}+18\mathbf{j})\,\mathrm{N}$
   **b**  **i**  $-60\,\mathrm{N\,m}$    **ii**  $-72\,\mathrm{N\,m}$    **iii**  $-162\,\mathrm{N\,m}$

10  **a**  Forces $-2\mathbf{i}\,\mathrm{N},\,5\mathbf{j}\,\mathrm{N},\,(-1.6\mathbf{i}+1.2\mathbf{j})\,\mathrm{N}$,
      $(3.46\mathbf{i}+2\mathbf{j})\,\mathrm{N}$; Resultant $(-0.136\mathbf{i}+8.2\mathbf{j})\,\mathrm{N}$
   **b**  $20.4\,\mathrm{N\,m}$      **c**  $x=2.49$

## Exercise 7.3

1  **a**  No – non-zero resultant ($-8\,\mathrm{N}$)
   **b**  No – non-zero moment (12 Nm about every point)
   **c**  Yes – resultant and moments zero
   **d**  Yes – resultant and moments zero
   **e**  Yes – resultant and moments zero
   **f**  No – non-zero moment ($-8\,\mathrm{N\,m}$ about every point)

2  $R=8,\,x=1$

3  $46g\,\mathrm{N},\,64g\,\mathrm{N}$

4  **a**  $50\,\mathrm{N},\,150\,\mathrm{N}$      **b**  $0.2\,\mathrm{m}$

5  **a**  At $A$: $147-2.45m\,\mathrm{N}$, at $C$: $49+12.25m$
   **b**  $m\leqslant 60$

6  **a**  $1\,\mathrm{m}$      **b**  $2.5\,\mathrm{m}$

7  $15\leqslant m\leqslant 600$

8  **a**  $1.5\,\mathrm{m}$      **b**  $100\,\mathrm{N}$

9  $a<x<10a$

10  $\dfrac{W(a+c-2b)}{2c},\,\dfrac{W(2b+c-a)}{2c}$

11  **a**  $2\frac{2}{3}\,\mathrm{m}$      **b**  $12.5\,\mathrm{kg}$

13  **a**  $2.75\,\mathrm{m}$

14  $6\,\mathrm{kg},\,\frac{8}{15}\,\mathrm{m}$

15  **a**  $y=\dfrac{x+8}{5}$      **b**  $0<x<2$, so $1.6<y<2$

16  $m_2-3m_1$

## Review 7

1  **a**  $22\,\mathrm{N\,m}$      **b**  $2.53\,\mathrm{N\,m}$    **c**  $-14.3\,\mathrm{N\,m}$

2  $2\frac{1}{4}\,\mathrm{N},\,7\frac{2}{3}\,\mathrm{N}$

3  **a**  $28.8\,\mathrm{N\,m}$      **b**  $-10.9\,\mathrm{N\,m}$

4  **a**  $10\,\mathrm{N\,m}$      **b**  $-11\,\mathrm{N\,m}$

5  $48\,\mathrm{N\,m}$

6  $4.75g\,\mathrm{N},\,3.25g\,\mathrm{N}$

7  No more than $\frac{3}{7}\,\mathrm{m}$ from centre

8  **a**  $180\,\mathrm{N},\,310\,\mathrm{N}$      **b**  $m=120$

9  **a**  $1\,\mathrm{m}$
   **b**  No need to allow for size of children or positions
      of their centres of gravity

10  **a**  $10\,\mathrm{kg}$      **b**  $2.4\,\mathrm{m}$
   **c**  **i**  weight acts at centre of plank
      **ii**  plank does not bend
      **iii**  no need to allow for size or centres of gravity
         of Jack and Jill

## Revision 2

1  **a**  $(-5\mathbf{i}+12\mathbf{j})\,\mathrm{m\,s^{-2}}$    **b**  $5.2\,\mathrm{N}$

2  **a**  $22.6\,\mathrm{N}$      **b**  $23.8\,\mathrm{N}$

3  **a**  $25.2\,\mathrm{N}$      **b**  $43.6\,\mathrm{N}$
   **c**  **i**  mass of string insignificant compared to that
      of particles
      **ii**  tension same at $A$ and $B$

4  **a**  $3\,\mathrm{m\,s^{-2}}$      **b**  $14.8\,\mathrm{m\,s^{-1}}$    **d**  $3.07\,\mathrm{s}$

5  **b**  $16\,\mathrm{N}$
   **c**  both particles have same acceleration   **e**  $20\,\mathrm{m\,s^{-1}}$

6  $30g\,\mathrm{N},\,40g\,\mathrm{N}$

7  **a**  $60g\,\mathrm{N}$      **b**  $1.25\,\mathrm{m}$

8  **b**  $200x+W=675$    **c**  $x=3.1\,\mathrm{m},\,W=55\,\mathrm{N}$

9  **a**  $1.3\,\mathrm{m\,s^{-1}},\,2.2\,\mathrm{m\,s^{-1}}$    **b**  $22.5\,\mathrm{s}$

10  **a**  $2.43\,\mathrm{m\,s^{-2}}$      **b**  $9.74\,\mathrm{m\,s^{-1}}$

11  **a**  **i**  $0.4\,\mathrm{m\,s^{-2}}$    **ii**  $280\,\mathrm{N}$
   **b**  **i**  $2200\,\mathrm{N}$    **ii**  $400\,\mathrm{N}$    **iii**  thrust

12  **a**  $1.25\,\mathrm{m\,s^{-1}}$ in direction of $A$'s original motion
   **b**  $5250\,\mathrm{N\,s}$    **c**  trucks are particles    **d**  $12.5\,\mathrm{m}$

M1

## Vectors and scalars

A vector quantity has magnitude and direction.

e.g. Displacement, velocity, acceleration, force.

A scalar quantity has magnitude only.

e.g. Distance, speed, mass.

The vector shown is expressed in component form as

$$\overrightarrow{OP} = r\cos\theta\,\mathbf{i} + r\sin\theta\,\mathbf{j}$$

where $\mathbf{i}$ and $\mathbf{j}$ are the unit vectors in the $x$- and $y$-directions.

The magnitude and direction are given by

$$|\overrightarrow{OP}| = r = \sqrt{x^2 + y^2} \quad \text{and} \quad \tan\theta = \frac{y}{x}$$

## Motion in a straight line

Velocity is rate of change of displacement.

For constant (uniform) velocity:

$$\text{velocity} = \frac{\text{change of displacement}}{\text{time}}$$

For non-uniform velocity this gives the average velocity.

Speed is rate of change of distance.

For constant (uniform) speed:

$$\text{speed} = \frac{\text{distance travelled}}{\text{time}}$$

For non-uniform speed this gives the average speed.

Acceleration is rate of change of velocity.

For constant (uniform) acceleration:

$$\text{acceleration} = \frac{\text{change of velocity}}{\text{time}}$$

For non-uniform acceleration this gives the average acceleration.

The acceleration due to gravity is $g$. Near the surface of the Earth $g \approx 9.8 \text{ ms}^{-2}$.

The properties of motion graphs are:

|  | Displacement-time graph | Velocity-time graph | Acceleration-time graph |
|---|---|---|---|
| Gradient gives: | Velocity | Acceleration | – |
| Area gives: | – | Displacement | Change of velocity |

For uniform acceleration, the displacement ($s$), initial velocity ($u$), final velocity ($v$), acceleration ($a$) and time ($t$) are connected by five equations:

$$v = u + at$$
$$s = ut + \tfrac{1}{2}at^2$$
$$v^2 = u^2 + 2as$$
$$s = \tfrac{1}{2}(u + v)t$$
$$s = vt - \tfrac{1}{2}at^2$$

## Forces

The unit of force is the newton (N).

The weight, $W$, of an object of mass $m$ is the force caused by gravity. $W = mg$ N.

A normal reaction force occurs when surfaces are in contact. It acts at right angles to the plane of contact.

A friction force occurs when one surface is sliding or attempting to slide over another. It acts in the direction opposite to the motion.

A tension force occurs in a string or rod when external forces are tending to stretch it. For a light string/rod the tension is the same at every point.

A thrust (compression) force occurs in a rod when external forces are tending to compress it. For a light rod the thrust is the same at every point.

## Equilibrium

A particle is in equilibrium if it is stationary or travelling with constant velocity. The forces acting on a particle in equilibrium can be represented by a closed polygon of forces. For a three-force problem this is a triangle of forces.

For forces in equilibrium the total component in any chosen direction is zero.

## Friction

The coefficient of friction ($\mu$) is a constant relating the friction force and the normal reaction for a given pair of surfaces in contact.

If $F$ is the friction force, $R$ is the normal reaction and the object is at rest in equilibrium, then $F \leqslant \mu R$. If $F = \mu R$, the object is in limiting equilibrium.

If the object is moving then $F = \mu R$.

## Newton's laws

First law Every object remains at rest or moves with constant velocity unless subject to a resultant external force.

M1

Second law  The acceleration is in the direction of the force. The magnitude of the force is proportional to the magnitude of the acceleration and the mass of the object.

Using standard units, the second law is equivalent to $\mathbf{F} = m\mathbf{a}$. This is the equation of motion of the object.

Third law  For every action there is an equal and opposite reaction.

## Impulse and momentum

A force $\mathbf{F}$ N applied for a time $t$ s generates an impulse $\mathbf{F}t$ Ns.

A body of mass $m$ kg travelling with velocity $\mathbf{v}$ ms$^{-1}$ has momentum $m\mathbf{v}$ Ns.

They are related by impulse = change of momentum.

The principle of conservation of linear momentum states that the total momentum of a system in a particular direction remains constant unless an external force is applied in that direction.

## Moments

The turning effect of a force about a point is called its moment.

If a force $\mathbf{F}$ N acts at a perpendicular distance $d$ m from a point $A$, the magnitude of its moment about $A$ is $|\mathbf{F}|d$ Nm.

Conventionally, anticlockwise moments are positive, clockwise negative.

A system of parallel forces is in equilibrium if the resultant of the forces is zero and the sum of their moments about any point is zero.

M1

M1